ROUTLEDGE LIBRARY EDITIONS:
ETHICS

Volume 45

THE DEFINITION OF MORALITY

THE DEFINITION OF MORALITY

Edited by
G. WALLACE AND A. D. M. WALKER

Routledge
Taylor & Francis Group
LONDON AND NEW YORK

First published in 1970 by Methuen & Co Ltd

This edition first published in 2021
by Routledge
2 Park Square, Milton Park, Abingdon, Oxon OX14 4RN

and by Routledge
605 Third Avenue, New York, NY 10017

Routledge is an imprint of the Taylor & Francis Group, an informa business

British Library Cataloguing in Publication Data
A catalogue record for this book is available from the British Library

ISBN 13: 978-0-367-85624-3 (set)
ISBN 13: 978-0-367-50952-1 (hbk) (Volume 45)

Publisher's Note
The publisher has gone to great lengths to ensure the quality of this reprint but points out that some imperfections in the original copies may be apparent.

Disclaimer
The publisher has made every effort to trace copyright holders and would welcome correspondence from those they have been unable to trace.

The Definition of Morality

edited with an introduction by
G. WALLACE *and*
A. D. M. WALKER

METHUEN & CO LTD
11 NEW FETTER LANE LONDON EC4

First published 1970 by Methuen & Co Ltd

Printed in Great Britain by
The Camelot Press Ltd, London & Southampton
416 15110 8 hardboard
416 18040 x paperback

Distributed in the U.S.A.
by Barnes & Noble Inc.

CONTENTS

ACKNOWLEDGEMENTS

We are grateful to our colleagues Professor Alan White, Mr Richard Swinburne and Miss Ann Tuxill for their criticisms of an earlier draft of the Introduction; and to Mrs Lorna Haysom, Mrs Janet Middleton and Miss Judy Lunn for their help with the typing.

It is probably superfluous to add that the papers we reprint here do not necessarily represent the present opinions of their authors.

G. W.

A. D. M. W.

August 1969 University of Hull

Introduction

G. WALLACE AND A. D. M. WALKER

I

There has recently been much discussion among moral philosophers about the definition of 'morality'. What, they have asked, are the essential and defining characteristics of morality and moral issues, the characteristics which distinguish them from, say, the law and legal issues or etiquette and matters of good taste? Again, how do moral views differ from political views, moral beliefs from religious beliefs, moral judgements from aesthetic judgements? It is this question, or rather this group of questions, to which the papers brought together here are, in different ways, relevant.

To emphasize the need for such an investigation is hardly necessary. Alike in philosophical and non-philosophical discussion the words 'morality' and 'moral' are of frequent occurrence and it seems that in both areas confusion has arisen as a result of uncertainty or unclarity about their meaning.[1] Indeed, what is surprising is that detailed and self-conscious examination of the concept, or concepts, involved should not have been undertaken earlier.

There is, however, a general objection which some philosophers, especially those influenced by Wittgenstein, would make

[1] See, for example, G. J. Warnock *Contemporary Moral Philosophy*, p. 52, where it is suggested that Hume's and Kant's analyses of morality are not in fact analyses of the same thing. Consider also the discussion on 'Prudence' by J. D. Mabbott and H. J. N. Horsburgh (in *Proceedings of the Aristotelian Society*, Supplementary Volume, 1962); the disagreement here between Mabbott and Horsburgh over the extent to which prudence is a moral virtue is in part attributable to differing conceptions of what it is for a virtue to be a *moral* virtue. See also G. E. M. Anscombe, 'Modern Moral Philosophy', *Philosophy*, 1958 (reprinted here).

to the kind of investigation we have in mind. They would say that the search for definitions (that is, the search for conditions necessary and sufficient for the application of a word[1]) is philosophically useful only when all the things to which a word applies share an essence or possess one or more common properties. And has not Wittgenstein shown that what we find is not an essence or a number of common features but a set of family resemblances?[2] We cannot, for example, discover any characteristic common to all the activities called games; yet, just as the members of a family may resemble one another without there being any single feature (such as the family chin or nose) which they all possess, so too the various things to which a word such as 'game' applies resemble one another without sharing any common property or essence. Clearly, to look for a definition of 'game' (in the sense of a statement of the necessary and sufficient conditions of something's being a game) is futile; similarly, the Wittgensteinian would conclude, it is misguided to look for definitions of the words 'moral' and 'morality'.

The simple reply to this is that we need to consider whether what Wittgenstein says of 'game' is true of 'moral' and 'morality'.[3]

[1] For the sake of beginners, we should perhaps briefly explain the terms 'necessary condition' and 'sufficient condition', especially as we shall use them a great deal in the second section.

A condition C is a *necessary* condition of something's being an X if that thing cannot be an X without condition C's being satisfied; that is to say, condition C must be satisfied if the thing is to be an X. (Clearly, though, condition C can be satisfied by many things which are not X's.)

A condition C is a *sufficient* condition of something's being an X if its satisfaction guarantees that that thing is an X; that is to say, if condition C is satisfied, the thing must be an X. (To say that C is a sufficient condition of something's being an X is *not* to say that *all* X's satisfy this condition. C may be very specific and thus satisfied by comparatively few X's.)

A condition C is a *necessary and sufficient* condition of something's being an X if nothing can be an X without condition C's being satisfied and if also condition C's being satisfied guarantees that that thing is an X.

It follows that in attempting a definition of 'X' we should look for necessary conditions which are fairly restricted and sufficient conditions which are not too specific. Ideally we are seeking a condition which is necessary and sufficient.

[2] Ludwig Wittgenstein, *The Blue Book*, p. 17 (Basil Blackwell, 1958); *Philosophical Investigations*, paras. 66 ff. (Basil Blackwell, 1953).

[3] Further, what Wittgenstein says of 'game' is not beyond dispute. See H. Khatchadourian, 'Common Names and "Family Resemblances"', *Philosophy and Phenomonological Research*, 1957–8; M. Mandelbaum, 'Family Resemblances and

Plainly not all words are family resemblance words; some words *are* applied to things or actions all of which *do* have something in common (e.g. 'triangle', 'walking'); so for any particular word we need to discover whether it is a family-resemblance word or not. Thus in the field we are concerned with, one can start by looking systematically for the necessary and sufficient conditions of, for instance, a principle's being a moral principle; and if at the end of this and similar searches we cannot find any such conditions, then we should take seriously the claim that 'moral' and 'morality' are family-resemblance words. For, apart from the support which would be so given to the Wittgensteinian thesis, we should perhaps be in a better position to discover which similarities and relations constitute family resemblances among the things referred to by means of 'moral' and 'morality'. This is because in looking for necessary and sufficient conditions we are likely to be struck by those features which, though perhaps inadequate as necessary conditions or sufficient conditions, nevertheless are important enough to be or to indicate family resemblances.

The foregoing objection is, of course, quite general in the sense that it is intended to have applicability to attempts to define many words besides 'moral' and 'morality'. In the remainder of this first part of the Introduction we shall concentrate on some of the complexities and difficulties peculiar to the definitions of these two terms, before turning in the second part to the various accounts, sometimes very tentative and schematic, which have been offered.

In the first place, one must bear in mind the great variety of contexts in which the words 'moral' and 'morality' are used. It is a commonplace that sometimes 'moral' is a term of approval and is opposed to 'immoral' or 'morally wrong', and that sometimes it is a classificatory term and has as its contradictory 'non-moral'. But even as a classificatory term 'moral' occurs in a wide range of contexts. We speak, for example, of an individual's moral principles, moral views and moral convictions and may distinguish these from his religious principles, views and convictions.

Generalization Concerning the Arts', *American Philosophical Quarterly*, 1965; A. Manser, 'Games and Family Resemblances', *Philosophy*, 1967.

But 'moral' is equally at home when applied to nouns which, unlike 'principle', 'view' and 'conviction', cannot be qualified by a possessive: thus we speak of the moral point of view, moral reasons, moral considerations, and, rather differently, of the moral law (but not of *my* moral point of view and *your* moral law). Another interesting feature of our use of 'moral' is the way in which in one combination it contrasts with one set of epithets, whereas in another combination it contrasts with a different set: a moral *duty* can be contrasted with a legal duty (but not with an aesthetic duty); on the other hand, moral *considerations* may be opposed to legal, prudential or aesthetic considerations; and in 'moral *belief*' 'moral' has a rather different set of contraries – one cannot, for instance, speak of a legal or an aesthetic belief. (Whether 'moral' has a single sense in all these contexts, or whether the facts we have mentioned are sufficient to establish that the word has several different senses will not be discussed here; but the possibility that the word does have more than one sense cannot be excluded.)

The situation with 'morality' is roughly similar. At times the word refers to a certain kind of *conduct* and has as its opposite 'immorality' (as in 'His immorality scandalized the neighbourhood' and 'He required a high standard of morality from his children'). Perhaps more frequently – and in this sense the word has no opposite – it refers to a set or system of beliefs or rules about conduct; it is in this sense that we speak of Christian morality or the morality of fifth-century Greece. Yet again the word sometimes means 'the quality or fact of being moral'; thus we may discuss the morality of abortion or regicide.

That 'moral' and 'morality' occur in this way in a variety of contexts must have significance for anyone who would offer a definition of the terms, inasmuch as the definition he gives of 'moral' (or 'morality') in any one context should be capable of extension to cover the use of the word in other contexts in which it is used in the same sense; what he says about 'moral principle', for instance, should shed light on, and make good sense of, 'moral obligation', 'moral belief', and so forth.[1]

[1] Incidentally, it will be found that some of the definitions which have actually been offered go more happily with 'moral' and 'morality' in certain contexts and

What we have already said implies that we are concerned with concepts which have an everyday use and which are, therefore, expressed in everyday or ordinary language. Hence the adequacy of any proposed definition is to be settled at least in part by an appeal to ordinary language. But it is here that a notorious difficulty arises. For does this procedure not presuppose that in ordinary language 'moral' and 'morality' have a fairly precise and consistent use, and is it not the case that, in fact, these words have *no* precise and consistent use? C. H. Whiteley[1] argues for an affirmative answer to both these questions and maintains that the task of defining 'morality', conceived as the task of describing the meanings of 'moral' and 'morality' as these words occur in ordinary language, is futile. The test of a definition of 'moral' or 'morality' cannot be its correctness or incorrectness, its correspondence or lack of correspondence to ordinary usage, but must rather be how well it serves the purposes for which the definition is required.

But is it true that, as Whiteley contends, 'moral' and 'morality' in ordinary language have no precise and consistent use? To be sure, there is often disagreement about whether certain problems or principles are, or are not, moral. Professor R. M. Hare instances a man who is trying to decide whether to become an army officer or a stockbroker.[2] If, Hare asks, the man chooses to go into the army on the grounds that 'stockbroking is a sedentary and sordid occupation and the military life an active one, requiring courage and self-sacrifice', are we to say that he has been swayed by *moral* considerations – granted that 'it will not make a predictable difference to other people's interests' which decision he makes? To some people, as to Hare, it would be perfectly acceptable to speak of moral considerations, but to others the situation would be almost a paradigm case of a man's being influenced by other than moral considerations. The existence of such evidence – and cases of similar disagreement can be multiplied – is undeniable; but is it sufficient to justify Whiteley's

cannot readily be extended to cover the words in others. See A. MacIntyre, 'What morality is not', *Philosophy*, 1957 (reprinted here).

[1] 'On defining "moral" ', *Analysis*, 1959–60 (reprinted here).

[2] *Freedom and Reason*, p. 148.

contention that in ordinary language 'moral' and 'morality' have no precise and consistent use?

The only answer can be that it is not. An explanation of the disagreements lies in the possibility, already hinted at, that the words 'moral' and 'morality' have more than one sense. If, as seems likely, there are two or more distinct senses of 'moral' and 'morality', then it may be that the parties to such disputes are each using 'moral' consistently but in different senses. Such a possibility is not considered by Whiteley, but it is entertained by Hare, and is taken up by a number of contributors to the collection.[1] If there are several different senses of 'moral' and 'morality', the rules governing the use of the word in any one of these senses may yet be consistent and comparatively strict and precise. If so, the task of defining the words would, of course, be more complex than appeared at first sight, but not on that account impossible.

It is also worth mentioning another possibility. It may be that, in some of their uses, 'moral' and 'morality' *are* rather vague words, that there is a range of borderline cases where usage is fluid and there is room for dispute. Perhaps Whiteley is resting his case on this. But surely this fact, if it is one, is hardly very damaging, so far as concerns the definition of 'morality' – indeed much the same could be said of almost any word. Nor should the existence of disagreement over borderline cases blind us to the existence of universal agreement and consistency over the central cases. For every issue which is disputably a moral issue it is possible to produce one which is indisputably a moral issue.

Finally, having dealt with this objection and before turning to the various types of definition which have been offered, we should perhaps say a little more here about how the satisfactoriness or otherwise of these definitions is to be assessed. We have already said that an adequate definition must allow us to use the words 'moral' and 'morality' with good sense over the whole range of contexts in which we do use them. Earlier we mentioned the variety of nouns which may be qualified by the adjective 'moral',

[1] See for example, W.K. Frankena, 'The Concept of Morality', *University of Colorado Studies*, Series in Philosophy No. 3, 1967; P. F. Strawson, 'Social morality and individual ideal', *Philosophy*, 1961; N. Cooper, 'Two concepts of morality', *Philosophy*, 1966, and 'Morality and Importance', *Mind*, 1968.

and it will be appropriate to conclude this section by emphasizing further the great variety of ways in which the words occur. Since in the following section we shall deal mostly with moral principles, let us begin with them. We may accept, adopt, or reject a moral principle, or something *as* a moral principle; we may live up to, or fail to live up to, our moral principles; should we do the latter, we may blame, criticize, or castigate ourselves for so doing, or feel regret or remorse. Rather similarly, we may accept or reject that something is one of our moral duties or obligations; fulfil or fail to fulfil our moral duties and obligations, or (more rarely) go beyond what these duties and obligations require of us; again, praise and blame, justification and excuse may be offered as appropriate. In our relations with others we may advise or persuade, urge or exhort them to act in certain ways which we consider to be morally right (though equally, if less admirably, we may advise them *not* to act in those ways *even though* they are right); we may criticize them, or in other ways express our disapproval, when they fail to accept our advice, or, having accepted it, fail through some weakness to act on it. Even when a man has acted in accordance with his moral principles and has done what, in his view, was morally right, he is not immune from criticism; we may criticize a man although he thinks that he acted rightly. There is, lastly, a deliberative side to morality. We may decide whether to adopt a particular moral principle, discover whether we are under a moral obligation to act in a certain way, find out what would be the right thing to do in such-and-such circumstances. Moral considerations must be weighed, moral scruples heeded or ignored, moral doubts dissolved or suppressed. We may call in the help of friends, whose moral views we may on reflection consider sound or unsound, enlightened or unenlightened, well thought out or superficial, and so on. We may ask, even, why we should give any weight at all to moral considerations or whether they should always be given overriding weight; and in doing so we shall embark on the task of justifying morality.[1]

[1] See further P. F. Strawson, 'Social Morality and Individual Ideal', *Philosophy*, 1961 (reprinted here); and B. A. O. Williams, 'Ethical Consistency', *Proceedings of the Aristotelian Society*, Supplementary volume, 1965, where attention is given to other, sometimes neglected aspects of morality.

II

We turn now to the definitions of 'moral' and 'morality' which philosophers have actually offered. But first a word of explanation. Throughout this section we are concerned not so much with particular definitions proposed by particular philosophers as with *types* of definition. Accordingly, many of our criticisms are to be understood not as conclusive objections to particular definitions, but rather as obstacles, perhaps superable, perhaps not, in the way of *particular kinds of approach* to the problem.

(i) *Moral rules and principles are to be defined by reference to their universalizability*

Largely owing to the immense influence of Professor Hare's work[1] the concept of universalizability has recently received a great deal of attention from moral philosophers. Hare's claim that moral principles are universalizable is, it would seem, to be understood in the following way: if I maintain that morally I ought to do X, then I am committed to maintaining that morally anyone else ought to do X unless there are relevant differences between the other person and myself and/or between his situation and mine.[2]

One crucial dispute concerns whether the criterion of universalizability is purely formal or substantive: is it logic (that is, the meaning of the word 'moral') which requires that I universalize my moral principles or is the principle that I should do so itself a substantive moral principle? And included in this volume are papers by Alasdair MacIntyre and W.K. Frankena[3] which bear directly on the status of the universalizability criterion.

As a necessary condition of a principle's being a moral principle, the criterion of universalizability obviously has much to be said

[1] R.M. Hare, *Language of Morals*, Oxford, 1952; *Freedom and Reason*, Oxford, 1963. Hare's work derives in part from that of Kant; see especially *Groundwork of the Metaphysic of Morals*.

[2] Hare usually states his theory not in terms of principles but in terms of judgements. This does not, however, affect what we have to say.

[3] A.C. MacIntyre, 'What morality is not', *Philosophy*, 1957; W.K. Frankena, 'MacIntyre on defining "morality"', *Philosophy*, 1959: See also T.L.S. Sprigge, 'Definition of a Moral Judgement', *Philosophy*, 1964 (reprinted here).

in its favour. Equally obviously, however, it will not do as a sufficient condition because many other, indeed perhaps all other, sorts of principles are universalizable. Consider, for instance, the principle that in oil-painting bright colours should be painted in last; this is universalizable but would not be classified as a moral principle.

(ii) *Moral rules and principles are to be defined by reference to their prescriptivity*

Like the claim that moral rules and principles are universalizable, the claim that they are prescriptive may be understood in several ways.[1]

It may be understood, first, as advancing the comparatively weak thesis that moral rules and principles are prescriptive in the sense that they are, in some way left unspecified, action-guiding. This would explain, for example, our finding odd the remark of a man who claimed to hold the principle that he ought to do X and yet, while agreeing he had many opportunities to do X, never did so. Understood thus the claim is in roughly the same position as the preceding claim that moral rules and principles are universalizable: it could plausibly be said to state a necessary, but could hardly be said to state a sufficient, condition of a rule or principle's being a moral rule or principle; many other sorts of rule and principle (for example, 'One ought not to swim immediately after a meal') are action-guiding.

The claim may, however, take a stronger and, we think, less plausible form. It then seeks to explain – and this is the form the claim takes in Hare's work – how, or in what way, rules and principles are action-guiding. It argues that their action-guiding force derives from the fact that they entail imperatives: my acceptance of the principle 'One ought to do X' commits me to accepting the imperative 'Let me do X'; and my acceptance of the imperative commits me in turn to doing X in the appropriate circumstances.[2]

This Harean form of prescriptivism gives rise to complex and far-reaching problems. Among these is the notorious problem of

[1] Compare G. J. Warnock, *Contemporary Moral Philosophy*, Ch. IV.

[2] R. M. Hare, *Language of Morals*, and *Freedom and Reason*, *passim*.

weakness of the will: how is the prescriptivist to account for the fact that a man may fail to live up to his principles? Hare is forced to explain the possibility in one of a severely restricted number of ways. On his view, the man who maintains that he ought to do X but does not, either does not really hold that he ought to do X or else is prevented in some way from doing X. Thus, if Hare were right, it would be logically impossible that a man who could, on a particular occasion, live up to a sincerely held principle should, on that occasion, fail to do so; from which it would follow that we ought to be more inflexible than we are in deciding whether a man does or does not sincerely accept a given principle. As it is, a modest amount of backsliding does not cast doubt on his sincerity.[1]

(Incidentally, it may be suggested that while neither the criterion of universalizability nor the criterion of prescriptivity, taken singly, provides a sufficient condition of a rule or principle's being a moral rule or principle, taken jointly they *are* adequate. Unfortunately this will not do either: other sorts of rules and principles possess both the characteristics of universalizability and prescriptivity.[2])

(iii) *Moral rules and principles are to be defined by reference to the fact that they override other sorts of rules and principles*[3]

Many philosophers have been attracted by the idea of defining, or partly defining, 'morality' in terms of the relationships which moral rules and principles have, or are believed to have, to other kinds of rules and principles. Such a programme is attractive because we do tend to think that moral rules and principles have a status which other rules either lack or need not have. The thesis that moral rules override other kinds of rules may be seen, therefore, as an attempt to define 'moral' and 'morality' in terms of this status. Moral rules and principles, on this view, have a status which other rules and principles lack because the former override the latter.

[1] For a discussion of this and some of the other difficulties in Hare's prescriptivism see C. C. W. Taylor, 'Critical notice of Hare's *Freedom and Reason*', *Mind*, 1965, and A. C. MacIntyre, 'What morality is not', *Philosophy*, 1957 (both reprinted here).

[2] See, for example, Hare, *Freedom and Reason*, Ch. 9.

[3] Variants of this thesis are to be found in Hare, op. cit., p. 169 and in N. Cooper, 'Morality and Importance', *Mind*, 1968 (reprinted here).

What is being claimed when it is said that moral rules override other kinds of rules? To this question there are two radically different answers. First, what might be meant is that when people are confronted with a choice between acting in accordance with a moral rule and acting in accordance with a non-moral rule, they always act in accordance with the moral rule. Or, secondly, it might mean that people who accept moral rules believe that they *ought* to act in accordance with those rules when they conflict with other kinds of rules.

On the first interpretation it becomes impossible for a person or society to adopt a moral rule and fail consistently to act in accordance with it when it conflicts with other kinds of rules; if the rule is not conformed to, it will not override other kinds of rules and will not be a moral rule. But obviously men sometimes do fail in this way to act in accordance with moral rules. It looks, therefore, as though the second interpretation, which is not open to the same objection, has more chance of succeeding.

But if this thesis were correct we should be forced to say that a man who, while recognizing that he has a duty to provide for his wife and family, believes that he has an overriding duty to paint, must have or believe he has a moral duty to paint. It is true that sociologists and anthropologists sometimes count as part of a society's morality those rules which are taken to be of overriding importance. But it is far from clear that in the example we are considering the meaning of 'moral' forces us to say that the painter believes he has a moral duty to paint. Further, if the thesis under consideration were true, we should expect such remarks as 'Other kinds of rules ought sometimes to override moral rules' to be self-contradictory. It may be that this is so, but as far as we know, it has not been established.

(iv) *Moral rules and principles are to be defined by reference to the fact that they have an importance which other rules and principles do not have or need not have*[1]

[1] Variants of this thesis are to be found in N. Cooper, 'Morality and Importance', *Mind*, 1968 (reprinted here); S. Hampshire, *Thought and Action*, p. 249; H. L. A. Hart, *The Concept of Law*, Ch. 8. It is sometimes offered to elucidate the difference between moral rules and, on the other hand, legal rules and rules of etiquette.

Although it is easy to confuse this thesis, or a variety of it, with some variety of the preceding thesis, they are in fact quite distinct. If rule X overrides rule Y, it does not follow that rule X is, or is held to be, more important than rule Y. Nor from the fact that rule X is, or is held to be, an important rule to observe does it follow that rule X overrides other – still less all other – kinds of rules.

As with the last thesis, we can distinguish two different ways in which the notion of importance could be useful in an attempt to define 'moral'. A rule or principle, it might be said, is a moral rule or principle if it is (1) held to be important by its holders or (2) is important. To the best of our knowledge no one has maintained that either interpretation provides a sufficient condition of a rule or principle's being a moral rule or principle, so we shall confine ourselves to a discussion of these interpretations taken as stating necessary conditions.

How are we to decide whether it is true that a necessary condition of a rule's being a moral rule is that it is, or is held to be, important? First we must ask what is the basis of comparison in accordance with which moral rules will do well and – at least some – other rules not so well. Normally when we compare A and B and conclude that A is more important than B, we are able to specify the respects in which A is, or is held to be, more important than B. We can speak of things as being morally important or important from the, or a, moral point of view; but we also talk of legal, political, aesthetic and economic importance. Now if we maintain that moral rules are those which it is morally important to keep, we have plainly shed no light on the meaning of 'moral' because we have tried to explain 'moral' in terms of 'morally'. If, therefore, either of the theses which seek to define 'moral' in terms of importance is to be acceptable there must be some basis of comparison other than that of moral importance. One possibility is that the importance of moral rules is to be located in the fact that if certain moral rules (for example, those concerned with the use of violence and the taking of life) were to be widely disregarded social chaos would almost certainly ensue. Obviously, though, moral rules are of *differing* importance in this respect and it is not true of *all* moral rules that widespread disregard of them would lead, or is held to lead, to disastrous social consequences. At best, then, we can say that, *as a kind*, they are important when compared with some other kinds of rules. (This, however, does not seem to be the case when they are compared with legal rules.)

Thus it looks as though it would be better to argue for the importance of moral rules in some other way. One such way would be to show that the acceptance of a moral rule necessarily commits one to being prepared, should the occasion arise, to give second place to one's own wants, desires and interests. The importance of moral rules would then be exhibited by reference to the strong pressures which need to be overcome in keeping to the rules. Their importance would be shown in terms of how much a man is prepared to give up in order to act in accordance with them.

However, it could be objected that even if moral rules *are* shown to be important in this way it is the notion of accepting a rule rather than that of morality which gives plausibility to the claim; or, to put the same point slightly differently, that what we have shed light on is not the concept of morality but that of accepting a rule (in the sense of acting in accordance with a rule). After all, it might be said, if one accepts, that is, acts in accordance with, the rules of cricket or football, one must be prepared to do things which one would rather not do. It would thus seem that this variant of the importance thesis is rather insecurely based on the fragment of truth that conformity to some moral rules does involve most people in overcoming strong temptations to act in other ways.

The previous paragraph might be thought to beg the question unfairly. For a definition of the type we are considering asserts that part of what it *means* to call a rule a moral rule is that it should be accepted in this way, that is, acted on in the face of strong desires or inclinations to the contrary. Now what we have said is, admittedly, open to this objection; but then we should reply that to call a rule a moral rule manifestly does not mean this. If it did, we should be faced with, among other things, the paradox that, should no one ever feel a strong urge to kill others, the rule that one ought not to kill would cease to be a moral rule.

(v) *Moral rules and principles are to be defined by reference to the particular forms of sanctions associated with them*[1]

Under this heading we can for convenience consider two radically different kinds of sanctions which, it has been said, attach to moral rules and principles. First, moral rules, it is said, are accompanied by specific forms of social pressures; the person who breaks the rules invites hostility, contempt, unfriendliness and perhaps ostracism. Secondly, it is said, the person who fails to act in accordance with a moral rule or his own moral principles experiences feelings of guilt, shame or remorse. (When social scientists distinguish between external and internal sanctions they sometimes have in mind these two different kinds of sanctions.)

Certainly it is true that if a man ignores or flouts the moral rules of his society, he may be treated with contempt or unfriendliness. It often happens also that we feel guilt and remorse when we fail to live up to our own moral standards. What is not so clear is whether these features of moral life tell us anything about the meaning of 'moral' or 'morality'.

There are fairly obvious objections to taking the presence of either of these forms of sanction as a sufficient condition of a rule or principle's being a moral one. One such objection is provided by those types of neurosis which involve misplaced guilt feelings. Some neurotic patients, for example, suffer great mental anguish if they do not constantly wash their hands; but surely it is implausible to say that such patients have, or think they have, a moral duty constantly to wash their hands; nor should we accept that because they feel guilt and shame they have lapsed morally.

It is equally implausible to maintain that because I break an accepted rule in my society or social group and as a result am treated with contempt and unfriendliness, the rule must be a moral one. Some take rules of etiquette very seriously. Are we to say that when eating peas with a knife brings down on our heads the contempt and unfriendliness of such individuals, we have

[1] Variants of this thesis are to be found in H. L. A. Hart, *The Concept of Law*, Ch. 8; N. Cooper, 'Two concepts of morality', *Philosophy*, 1966 (reprinted here); and T. L. S. Sprigge, 'Definition of a moral judgement', *Philosophy*, 1964 (reprinted here).

done something morally wrong, in their eyes at least? Are their moral principles indistinguishable from their rules of etiquette? We shall have to say so if we maintain that the presence of contempt and unfriendliness is a sufficient condition of *moral* failure.

How does the presence of sanctions fare as a necessary condition of a rule or principle's being a moral rule or principle? It is important to remember here that one can question the morality of despising or ostracizing a person and that the morality of a society or an individual may contain a rule to the effect that it is always wrong to treat people in these ways. Perhaps the maxim 'Hate the sin but love the sinner' would be an instance. Further, with some individuals and some social groups we find a preference for persuasion and discussion rather than ostracism and contempt. It is, therefore, difficult to accept that the absence of the latter implies the absence of morality.

Finally, if a person never feels guilt or shame must he be either morally perfect or without moral principles altogether? If it is a necessary condition of one's having moral principles or of one's acceptance of moral rules that one feels guilt and shame on failing to act in accordance with them, it would seem that he must be one or the other. Some of the plausibility of this thesis arises from the same considerations as those on which many prescriptivist arguments rely, considerations relating to the action-guiding role which moral rules and judgements are said to play. Thus if a man says that torturing people is morally reprehensible and yet does so when not forced to, we shall begin to doubt his sincerity; but on discovering that he feels guilt or remorse for what he has done we may suspend these doubts. However, this shows at best that these feelings are good indices of sincerity; and the thesis we are considering requires more than this. If it is true, it must be impossible for a person to act contrary to a moral rule he accepts and not feel guilt, remorse or shame. And unfortunately the requirement cannot be met in some very obvious situations, namely those in which two or more of one's moral principles conflict. For example, if I hold that it is wrong both to tell lies and to endanger the lives of innocent people, I may feel called upon to tell lies; but provided I think I have done what morality requires of me in this situation there may be no reason why I should feel guilt or remorse.[1] This would suggest the thesis needs to be refined so that it allows for the possibility of conflict between moral rules or principles. One way in which this might be done is by relating the concepts of guilt and remorse to that of moral wrongness. The reformulated thesis would now maintain that it is a necessary condition of one's holding a particular action to be morally wrong that one feels guilt or remorse when one does the action.[2] This is, it seems, an attractive thesis but one not without difficulties. Its truth would make it impossible for a person to hold that what he has done was morally wrong while feeling neither guilt nor repentance. But is this the case? Suppose a child's parents are brutally murdered before his eyes; he vows he will take revenge and eventually does so without subsequent pangs of conscience or feelings of repentance. He says of the episode: 'I know [or believe] that what I did was wrong; nevertheless I would do it again.' Is the thesis we are examining so strong that we can simply conclude he does not really believe that what he did was wrong?

(vi) *Moral rules and principles are to be defined by reference to their content*

As with the earlier views, there are several variants of this general thesis.[3] It may be said first that a moral rule or principle is such because it mentions, as good or bad, right or wrong, certain kinds of actions. Secondly, it may be said that a rule or principle is a moral rule or principle if its holder justifies it by appealing to considerations of certain kinds. Thirdly, it may be said that moral rules and principles are to be characterized by reference to their being rules or principles with a certain kind of purpose (for example, the promotion of social harmony) or, slightly differently, by reference to their being rules or principles such that the consequences of an individual's accepting, or of everyone's

[1] Compare on this point B. A. O. Williams, 'Ethical Consistency', *Proceedings of the Aristotelian Society*, Supplementary volume, 1965.

[2] This variant would need further qualification in order to deal with actions done unintentionally, and so on.

[3] In the little discussion there has been, philosophers have not always kept these variants distinct. See, for instance, G. J. Warnock, *Contemporary Moral Philosophy*, pp. 54–5.

accepting them are, or are believed to be, of a certain kind.[1]

This classification is not intended to be exhaustive, nor should we like to suggest that the lines between the various species are at all clear-cut. It is not straightforwardly obvious, for example, whether Mrs Foot's early view[2] that a principle cannot be a moral principle unless a certain kind of *background* to the principle can be filled in is to be placed in the first, second or third of the categories we have distinguished or whether perhaps it is more satisfactorily placed in a new category outside these. Still, however rough and ready, the classification will do for our present purposes.

So long as we confine ourselves to schematic descriptions of these types of definition, there is little possibility for worthwhile discussion, since plainly the value of any definition along these lines depends to a very large extent on its detailed elaboration; it depends on precisely *what* moral rules and principles are defined to be about. Nonetheless, one important point can be made on this general level. It is easily seen that, whatever the specific form of the criterion proposed, it must be inadequate if meant as a *sufficient* condition of a rule or principle's being a moral rule or principle. It must be inadequate because, whatever the subject-matter – or the nature of the justifying considerations or the purpose – proposed for moral rules, it is possible that non-moral rules should be about this subject-matter – or have action in accordance with them justified in this way, or have this purpose. It would seem, in particular, to be especially difficult to separate moral rules from legal rules since legal rules could logically be about the same things, could have action in accordance with them justified in the same way, or could have the same purpose, as moral rules.[3]

[1] Views of the second kind have been advanced by, among others, P. R. Foot, 'Moral Arguments', *Mind*, 1958 (reprinted here); D. P. Gauthier, *Practical Reasoning*, Ch. 6; M. G. Singer, *Generalization in Ethics*, Ch. 6; and S. I. Benn and R. S. Peters *Social Principles and the Democratic State*, Ch. 2. In the third category must be placed the views of Kurt Baier, *The Moral Point of View*, Ch. 8 (reprinted here), and possibly those of J. S. Mill.

[2] 'When is a principle a moral principle?' *Proceedings of the Aristotelian Society*, 1954.

[3] It may be said that we could not have legal rules about what, for example, a man should think. But this, if true, seems only a practical, and not a logical, impossibility.

A criterion in terms of the content of moral rules and principles cannot, therefore, plausibly be advanced as a sufficient condition of a rule or principle's being a moral rule or principle; but then those who have advanced a criterion of this type have generally been careful not to seek such a role for it.[1] They have advanced it, rather, as a *necessary* condition of a rule or principle's being a moral rule or principle; and regarded in this light, it does indeed have considerable attractiveness. Our reluctance to describe such rules as 'Always wear a green tie' and 'Never use a ball-point pen' as moral rules does seem to stem from one, or perhaps all, of the related facts that they are not about the appropriate kind of actions, could hardly have action in accordance with them justified in the appropriate way, and could scarcely have the appropriate purpose.

Again, not much in the way of criticism can be undertaken in isolation from a study of specific detailed definitions – with which, incidentally, we have not as yet really been provided. But a few general comments are possible.

It would seem that definitions of the first type (according to which moral rules and principles are those which mention, as good or bad, right or wrong, certain specified kinds of action) are somewhat less promising than definitions of the second or third types. On this former type of definition it must be the case that certain kinds of action could never be the subject of moral rules and principles; but given, for example, the eccentricity of some human beliefs about the consequences of actions, it would be rash, at the very least, to make such a claim.[2] Not being vulnerable to this objection, definitions of the second and third types are in a stronger position – not that they too do not present difficulties of their own. With the second type of definition, which affirms that a necessary condition for a rule or principle to be a moral rule or principle is that its holder should justify action

[1] See P. R. Foot, *Proceedings of the Aristotelian Society*, Supplementary volume, 1954; and G. J. Warnock, *Contemporary Moral Philosophy*, pp. 57–8.

[2] Possibly, though, in an amended form this kind of thesis might fare better. In its new form it would assert that a necessary condition of a rule or principle's being a moral rule or principle is that the actions it enjoins or forbids are describable or can be seen in certain very general ways, e.g. as contributing to human well-being.

in accordance with it (or perhaps be prepared to justify such action) by appealing to considerations of certain kinds (for example, those to do with human well-being), we appear to commit ourselves to the view that an unreflective or untheoretical person, who simply accepts certain rules or principles without asking for reasons and who sees no need for, or is unable or unwilling to provide any reasons, cannot have *moral* rules or principles – a harsh, if not a paradoxical, commitment. (Provided, however, this thesis states merely what sorts of consideration must be appealed to *if* a justification is to be given, the objection loses its force.) Definitions of the remaining type are not so obviously open to this objection because they state that it is the *purpose* of the rules or the *consequences* of their acceptance – rather than the kind of justifying considerations – which place them within or outside the sphere of the moral. But they in turn bring their own drawbacks. In the first place, the notion of the purpose of a moral rule[1] can hardly be elucidated without reference to the considerations its holders use, or would use, to justify action in accordance with it, so that definitions in terms of the purpose of moral rules will after all suffer from weaknesses similar to those of the second type. If we appeal not to the purpose of moral rules but to the *putative consequences* of their acceptance, we are in a similar difficulty; for, conceivably, a man may have no views about what are the consequences of acting in accordance with his moral principles. Nor will it help to appeal to the *actual consequences* of their acceptance inasmuch as this will entail that moral rules and principles based on erroneous beliefs cannot be moral rules and principles after all. It might be said, for instance, that the consequence of the acceptance of moral rules and principles was the promotion of social harmony; but if so, would one not be forced to admit that moral rules and principles which were based on erroneous beliefs and in fact created social strife were actually not moral rules and principles? A definition along these lines *might* give an adequate account of 'moral' as a term of approval; but would quite clearly be inadequate with 'moral' as a classificatory term.

[1] It seems strange to talk of the purpose of a principle.

III

Our review of the various types of definition completed, it can be seen that none of the theses we have examined states a sufficient condition for a rule or principle's being a moral rule or principle, and that understood as stating a necessary condition, many of them present difficulties. Of course (as we have hinted), in a more sophisticated form one of these theses might provide a satisfactory definition, or so might a thesis which combined elements from several types. To explore these possibilities, however, is not proper matter for an introduction.

I

On defining 'moral'

C. H. WHITELEY

A good many recent philosophical papers have been concerned with discussing which principles, attitudes, problems, propositions can be properly counted as 'moral', which characteristics are 'essential' to what is moral as against what is not. These are questions of definition. But the philosophical problem is not that of giving a 'correct' definition, in the sense of one which accords with ordinary usage. For in ordinary usage the words 'moral' and 'morality' have no precise and consistent use. While there are many principles, attitudes, problems which everybody would agree in calling 'moral', and many others which everybody would agree in not calling 'moral', there are large numbers of doubtful cases, and no generally accepted criteria for drawing the line. Thus a reasonably exact definition must depart from usage to some extent. Such a definition should not be judged as correct or incorrect; it should be judged as suitable or unsuitable. But we cannot judge of its suitability unless we have some idea what purpose the definition is intended to serve, and in what contexts it is to be used. Different definitions may well be convenient in different contexts or for different purposes. I shall assume that any acceptable way of defining 'moral' and 'morality' must isolate something which plays a distinctive part in human life, and must enable us to distinguish matters of morality ('right and wrong') from matters of taste or preference, and matters of convenience or expediency, since it is with these matters that morality is

Reprinted from Analysis, *1959–60, pp. 141–4, by permission of the author and* Analysis.

usually contrasted. I shall suggest two ways in which this can be done, and two ways in which it cannot be done.

The first possible way is that suggested by the etymology of 'moral', 'ethical' and similar words. The morality of a community consists of those ways of behaviour which each member of the community is taught, bidden and encouraged to adopt by the other members.

Moral behaviour is behaviour in accordance with these recommended patterns, moral grounds are grounds derived from applying the accepted rules, moral issues are issues involving the required standards. In this concept of morality there is included the idea of a sanction: no rule is part of a community's morality if people can openly break that rule without incurring the hostility and disapproval of their neighbours – there may, but need not, be other penalties as well. There is also included the idea of *general* rules or standard patterns of conduct, incumbent upon all of given classes of persons in given classes of circumstances; for only general rules or standards can be taught, enjoined and systematically enforced. There is not included in this concept any reference to motive; morality consists in habitual voluntary conformity with the conventions, from whatever motive or motives this conformity arises.

This is a sociological or political concept. Morality so defined has a definite role in the life of societies, and one may study its functions, its development, its relationship with religion, economics, government, etc. The concept suffers from some inevitable vagueness. Disapproval is a matter of degree; there is no clear line of demarcation between cutting a man because he is a bounder, and avoiding him because he is an eccentric. Besides, there are always certain kinds of conduct which some members of a given community wish to insist on, while others are indifferent. For these reasons, the distinction between what is morally obligatory and what is morally optional – a matter of taste or convenience – in a society cannot be quite sharp. But in most societies the doubtful cases are relatively few.

My second possible way of defining 'moral' is from the point of view of the agent himself. My morality consists, not in what other people insist that I should do, but in what *I* insist that I should do. It is the content of my conscience. The peculiarity of conscience, or moral obligation, as against other motives for action, is, as Kant observed, that its imperatives are categorical. My morality comprises those actions which I think I ought to do regardless of inclination and regardless of personal advantage. A man who just does whatever he feels inclined to do has no morals, even if what he does is always right and good. The same is true of a man who always does what he thinks will be for his own greatest happiness. Morality in this sense does not necessarily involve the idea of external penalties: I can hold myself obliged to do something though I shall be none the worse off, even in my neighbours' estimation, if I do not do it. Nor does it necessarily involve the idea of general rules or principles. It is both logically and psychologically possible for me to think that I am morally obliged to do something without thinking that anybody else in a similar situation is or would be obliged to do it. The instances *par excellence* of categorical imperatives, the 'calls' or 'concerns' which must be obeyed before all else ('The word of the Lord came unto me, saying . . .') are individual obligations without general import.

This concept of morality is psychological. It defines a certain factor in the consciousness and conduct of individuals. It is a suitable concept for those who are concerned with moral endeavour, aspiration and struggle, with the nature, development and influence of the conscience. It offers a sharper distinction between moral and non-moral than the first concept. For the difference between what I feel obliged to do willy-nilly, and what I am prepared to abandon if I lose my taste for it or find it unprofitable, while not perfectly clear-cut, can be pretty nearly so; and the more self-consciously conscientious I become, the sharper the distinction.

The two concepts of the moral coincide over a wide range: that is, the things that people think they (categorically) ought to do are very largely the things that are enjoined on them by their neighbours. But the coincidence between conscience and convention is logically contingent. It is possible for a community to have rules which, though generally obeyed, are obeyed solely from self-interest or habit, so that no member of it has any 'inner' morality. And it is possible for a community of people to have well-developed consciences which do not support any established conventions. No human community fits either of these possibilities. But in all human communities there is a divergence between conscience and convention; and it is usually wider than it looks on the surface. What people in their hearts are devoted to may be something quite other than the conventionally 'moral'. Here the study of language may easily mislead us. For the way people talk is dominated by the conventions, and can mask the way they think, feel and behave. In the extreme case, the man who denounces 'morality' or 'conscience' may have quite strong convictions about the way he ought to behave.

'Morality' can thus be defined in two different ways, from a sociological and from a psychological point of view. Contemporary philosophers, whose approach is apt to be neither sociological nor psychological but linguistic, sometimes attempt to draw a distinction between moral and non-moral in terms of uses of language or kinds of reasoning or argument. I do not believe that any such distinction can be drawn.

There are no words or expressions, no uses of words or expressions, no types of proposition, which are distinctively moral or ethical. There are indeed words and expressions whose characteristic use is to evaluate or to recommend, and it is important that this use of language, which is prominent in talk about morals, should be distinguished from other uses, such as the descriptive or the straight imperative. But words like 'good', 'right', 'ought', which are used to evaluate and to recommend in the discussion of moral issues, are also used, without change of meaning, in the discussion of other matters. Conduct can be good, as meat can be good. The fact that I can approve of conduct on moral grounds (though not only on moral grounds) no more makes 'good' an ethical word than the fact that I can approve of meat on dietetic or gastronomic grounds makes it a dietetic or gastronomic word. If I say to you that you *ought* to sell your brewery shares, I can give as my reason either that profits from brewing are declining, or that it is immoral to profit from the debaucheries of your fellows, or both reasons at once. There are many other words (words like 'honest', 'sly', 'rash', 'heroic', 'strait-laced', 'discreet', 'boorish', 'unselfish', etc.) which play an essential part in the expression of moral appraisals, though linguistic philosophers have paid them very little attention. But they can equally well be used in the expression of non-moral appraisals, or of neutral descriptions. Thus if we define Ethics as the study of the language of morals we give it no effective definition; for morals has neither vocabulary nor idioms of its own. And the examination of the meanings of words which are common to moral and non-moral discourse can hardly be expected to shed much light on the peculiar characteristics of morality.

Similarly, there are no specifically moral kinds of argument. Arguments aimed at convincing a person that he is morally obliged to do or not to do something may be of a variety of logical types: deduction from agreed premises ('But that would be *dishonest*!'); analogy ('Isn't this just like what X did, which we all thought was so shocking?'); appeals to sentiment ('Think how unhappy it would make her!'); sheer brow-beating ('I don't know how you have the nerve to suggest it!'). Sometimes we argue from a general principle, sometimes we attend only to the case in hand. Sometimes we assume a pro-attitude; sometimes we set to work to evoke it. There is no type of reasoning or persuasion in place in moral contexts which is not equally in place in admittedly non-moral contexts. Thus the 'moral' is not a subdivision either of language or of logic.

2

What morality is not

ALASDAIR MACINTYRE

The central task to which contemporary moral philosophers have addressed themselves is that of listing the distinctive characteristics of moral utterances. In this paper I am concerned to propound an entirely negative thesis about these characteristics. It is widely held that it is of the essence of moral valuations that they are universalizable and prescriptive. This is the contention which I wish to deny. I shall proceed by first examining the thesis that moral judgements are necessarily and essentially universalizable and then the thesis that their distinctive function is a prescriptive one. But as the argument proceeds I shall be unable to separate the discussion of the latter thesis from that of the former.

I

Are moral judgements essentially and necessarily universalizable? The contention that they are is expressed in its most illuminating form in Mr R. M. Hare's paper on 'Universalizability' (*Proceedings of the Aristotelian Society*, 1954–5, pp. 295–312). Mr Hare borrows his terminology from Mr E. Gellner's paper on 'Logic and Ethics' (*P.A.S.*, 1954–5, pp. 157–78) where Mr Gellner distinguishes what he calls U-type and E-type valuations. A U-type valuation is an application of 'a rule wholly devoid of any personal reference, a rule containing merely predicates (descriptions) and logical terms' (p. 163). An E-type valuation is one

---

Reprinted from Philosophy, *1957, pp. 325–35, by permission of the author and* Philosophy.

containing some uneliminable personal reference. Hare's thesis is that moral judgements are U-type valuations. To give a reason for an action is not necessarily to commit oneself to such a valuation 'for I see no grounds in common language for confining the word "reason" to reasons involving U-type rules' (p. 298). But Hare goes on to say that his thesis 'is analytic in virtue of the meaning of the word "moral"'.

What this amounts to is made very plain in an imaginary conversation which Mr Hare constructs between a 'Kantian' and an 'Existentialist'. This runs as follows:

E.: 'You oughtn't to do that.'

K.: 'So you think that one oughtn't to do that kind of thing?'

E.: 'I think nothing of the kind; I say only that *you* oughtn't to do *that.*'

K.: 'Don't you even imply that a person like me in circumstances of this kind oughtn't to do that kind of thing when the other people involved are the sort of people that they are?'

E.: 'No: I say only that *you* oughtn't to do *that.*'

K.: 'Are you making a moral judgement?'

E.: 'Yes.'

K.: 'In that case I fail to understand your use of the word "moral".'

Mr Hare's comment on this is: 'Most of us would be as baffled as the "Kantian"; and indeed we should be hard put to it to think of *any* use of the word "ought", moral or non-moral, in which the "Existentialist's" remarks would be comprehensible. Had the "Existentialist" said "Don't do that", instead of "You oughtn't to do that", the objections of the "Kantian" could not have been made; this illustrates one of the main differences between "ought" and ordinary imperatives' (pp. 304–5).

The crux then of Hare's position is the contention that whenever anyone says 'I, you or he ought to do so-and-so,' they are thereby committing themselves to the maxim 'One ought to do so-and-so.' This commitment is embodied in the meaning of the word 'ought' in so far as 'ought' is used morally – and indeed, Hare seems to say, in non-moral uses of 'ought' also. But is this contention in fact correct? Consider the following example which is borrowed from Sartre (*L'Existentialisme est un Humanisme*, pp. 39–42). One of Sartre's pupils was confronted during the war with the alternatives of leaving France to join de Gaulle or staying to look after his mother. His brother had been killed in the German offensive in 1940 and his father was a collaborator. These circumstances had left him with a strong feeling that he was responsible as a patriot and they had left his mother in a state of almost complete dependence upon him. What should he do? Stay with his mother or escape to England? Sartre uses this problem in order to argue that there are no 'objective' criteria by which such a choice may be made. Part of the force of his argument is this. Someone faced with such a decision might choose either to stay or to go without attempting to legislate for anyone else in a similar position. He might decide what to do without being willing to allow that anyone else who chose differently was blameworthy. He might legitimately announce his choice by saying, 'I have decided that I ought to stay with my mother.' If he did so, his use of 'ought' would not express any appeal to a universalizable principle. It would not be a U-type valuation, but it would be a moral valuation.

Two points need to be made about this example. The first concerns the function of 'I ought to do so-and-so' when it is used to announce a decision in a case like that of Sartre's pupil. Its use is plainly to commit oneself, to allow that if I do not do what I say I ought to do, then I am blameworthy. It is a performatory use of 'I ought' in that its use makes one responsible for performing a particular action where before saying 'I ought' one could not have been held responsible for performing that action rather than some alternative one. To note this is to bring out the oddity in Hare's treatment of the 'Existentialist's' contribution to his dialogue. For in this non-universalizable sense of 'ought' one could never say 'You oughtn't' but only 'I oughtn't.' To say 'You oughtn't' and suppose that you had used 'ought' in this sense would be as odd as to say 'You promise' and suppose that thereby one had committed someone else to a promise.

Secondly, it might be argued that the very possibility of a problem such as that of Sartre's pupil presupposes the acceptance of certain universalizable maxims as moral principles. If Sartre's pupil had not accepted the maxims 'One ought to assist one's parents when they are in need' and 'One ought to assist one's country when it is in need' there would have been no problem. What is important is that the clash between two principles need not be resolved by reformulating one of the principles or formulating a third one. Certainly this clash could be so resolved. Sartre's pupil might have acted on the maxim 'Duties to one's parents always have precedence over duties to one's country'. Had he done so he would have legislated not only for his own but for all relevantly similar situations. But in order to make his own decision he does not need to so legislate. Now it seems to be a consequence of Hare's position that if the decision between principles is itself to be a moral decision it *must* itself rest upon the adoption of a universalizable maxim. This, in the light of Sartre's example, could only be defended by an *a priori* restriction on the use of the word 'moral'. Such a restriction, however, would not be merely a restriction upon our use of a word. For to adopt Hare's use of 'moral' would be to permit only one way of settling conflicts of principle (that of formulating a new principle or reformulating an old one) to be counted as genuinely a moral solution to a moral problem, while another way – that of the non-universalizable decision à la Sartre – would be ruled out from the sphere of morality. To do this is plainly to do more than to offer a descriptive analysis of the meaning of 'moral'. It is to draw a line around one area of moral utterance and behaviour and restrict the term to that area.

What one can conclude from this is twofold. First, not all, but only some, moral valuations are universalizable. What leads Hare to insist that all are is his exclusive concentration on moral rules. For rules, whether moral or non-moral, are normally universal in scope anyway, just because they are rules. As Mr Isaiah Berlin has written in another context, 'In so far as rules are general instructions to act or refrain from acting in certain ways, in specified circumstances, enjoined upon persons of a specified kind, they enjoin uniform behaviour in identical cases' ('Equality', in *P.A.S.*, 1955–6, p. 305). If this is so, then there is nothing specific to moral valuation in universalizability and in so far as moral valuations are not expressions of rules they are not universalizable. Secondly, the exceptions are not simply cases analogous to that of Sartre's pupil. A whole range of cases can be envisaged where moral valuations are not universalizable. At the one extreme would be those instances where in adopting a moral position someone consciously refrains from legislating for others, although they might have done so; where a man says, for example, 'I ought to abstain from participation in war, but I cannot criticize or condemn responsible non-pacifists,' but might have said, 'One ought to abstain from participation in war.' In such a case whether to make a universal or a merely personal judgement is itself a moral problem. The fact that a man might on moral grounds refuse to legislate for anyone other than himself (perhaps on the grounds that to do so would be moral arrogance) would by itself be enough to show that not all moral valuation is universalizable. Or rather that once again this thesis can only be maintained by an *a priori* and quite unjustifiable restriction upon the word 'moral'. In other words, a man might conduct his moral life without the concept of 'duty' and substitute for it the concept of 'my duty'. But such a private morality would still be a morality.

More commonly, however, non-universalizable judgements occur when a man finds that the concept of 'duty' has limits which render it useless in certain situations of moral perplexity. Such is the example of Sartre's pupil. And such are the cases at the other end of our scale where moral valuations must be non-universalizable, where it is logically impossible to universalize. This is the case with what the theologians call 'works of supererogation'. A work of supererogation is by definition not numbered among the normal duties of life. Those duties – such things as keeping one's promises and paying one's debts – are partly characterized by the fact that the maxims which enjoin them are universalizable. But there are a great many acts of moral worth which do not come within their scope: one may be virtuous in the sense in which virtue is demanded of everyone without being morally heroic. A moral hero, such as Captain Oates, is one who does more than duty demands. In the universalizable sense of 'ought' it does not therefore make sense to assert that Captain Oates did what he ought to have done. To say of a man that he did his duty in performing a work of supererogation is to contradict oneself. Yet a man may set himself the task of performing a work of supererogation and commit himself to it so that he will blame himself if he fails without finding such failure in the case of others blameworthy. Such a man might legitimately say, 'I have taken so-and-so as what I ought to do.' And here his valuation cannot, logically cannot, be universalized.

II

Crucial to the argument so far that universalizability is not a necessary characteristic of all moral valuation has been the distinction between first person and third person uses of moral valuation. Before the force of this distinction can be fully understood, however, it is necessary to inquire what the function of moral valuation may be. The argument of this section will be that there are a great variety of uses to which moral utterance may be put, none of which can claim the title of 'the' function of moral valuation. It will be useful to list some of the tasks which even so familiar a form of moral judgement as 'x ought to do y' may be set.

(i) The expression of indignation or other violent or mild emotion. This is characteristically a function of third person uses. 'He ought to put his foot down,' we say angrily, although we might hesitate to advise him by saying, 'You ought to put your foot down,' and if we did so we would be advising as well as, or even rather than, expressing our emotion. Clearly, of course, we might say 'He ought to put his foot down' without any kind of emotion and here we would presumably be prescribing a course of action. If we did this we would be committing ourselves to the advice 'You ought to put your foot down' even if we never did in fact utter this advice. This brings out already how the same form of words may be put to quite different use in moral utterance. Those emotive theorists who said that the function of moral utterance was to evince emotion would therefore have been correct if they had substituted the indefinite for the definite article.

(ii) The expression of commands or exhortations. This is of

course a second person use. As Stevenson pointed out, we may often say (to a child, for example) 'You ought not to do that,' meaning simply 'Don't do that' (*Ethics and Language*, p. 21). We may, of course, and often do so use 'ought' that 'You ought' has more than or other than imperatival force, but we need not do so. Hare in his dialogue allows that the 'Existentialist' would have spoken correctly had he said 'Don't do that' instead of 'You oughtn't to do that.' But he does not consider the possibility that the substitution of 'ought' for an imperative might be something other than a universalizing of the imperative. Yet clearly in ordinary usage the use of 'ought' might simply be an indication of the importance attached to the imperative, as it could be in the case of the command to the child.

(iii) The appraisal of actions. We can appraise equally the past, present and future actions of ourselves and others, whether in their absence or to their faces. This is therefore one of the most general uses of 'ought' whether in tense or in person. When 'ought' is used for purposes of appraisal, it differs from 'good' in that comparatives and superlatives are not available. 'He did what he ought to have done' or 'He failed to do what he ought to have done' are the only two verdicts available. So that appraisal by use of 'ought' is appraisal that implies a single standard. If you made this use central to your moral utterance you would produce a morality akin to that of the Stoics where to fall from good in the slightest degree is to fall into evil. ('A man drowns in six inches of water as easily as in six feet.') At this point it is important to note that this use of 'ought' is logically independent of the imperatival use. That is, there is no inconsistency in saying 'You ought to do this, but don't.' Those philosophers who have insisted on analysing moral utterance in terms of imperatives would be forced to interpret this as meaning 'Do this (Let anyone in this sort of situation do this kind of thing), but don't,' which would be as nonsensical as any utterance of the form $p. \sim p$. But where the 'ought' expresses an appraisal there is no inconsistency. For however morally reprehensible it may be, there is no inconsistency in pointing out what the moral appraisal of an action would be and then suggesting that one act otherwise. Some writers have attempted to argue against this by interpreting

'You ought to do this, but don't' as meaning 'In this society most people would consider that you ought to do this, but I think you ought to do that.' But while this might be what would be meant, it need not be. A man might commit himself to a certain moral appraisal but not use it as a guide for action – 'This in the light of morality is how your action would be appraised: but don't follow the guidance of morality.'

(iv) The giving of advice. This is a genuinely prescriptive function of moral utterance. It is also one in which genuinely universalizable maxims are employed. For when we advise someone to undertake a certain course of action we do so in virtue of certain characteristics of their situation and certain characteristics of the recommended course of action. But while this might seem to be a use of 'ought' which accords admirably with Hare's analysis, there is one point that stands out for notice; namely, that the giving of advice is always a question of second person utterance. So that while 'You ought to do so-and-so' may express a universalizable prescription when it is offered as advice, clearly 'I ought to do so-and-so' cannot function in the same way. For one cannot advise oneself.

(v) Persuasion. So much has been said so well on this subject by Stevenson, that the only necessary comment is to point out that the present list of ways in which 'ought' can be used merely brings out the error of offering an analysis of 'ought' which restricts it to one of its possible uses.

(vi) The expression of one's own principles. This is the most characteristic first person use of 'ought'. But I do not think many people say 'I ought to do so-and-so' very often, and when they do, it is usually, I suspect, 'I ought to do so-and-so, but . . .' or 'I don't know what I ought to do over so-and-so.' In other words 'I ought' is used to express doubt and perplexity as well as and indeed perhaps as much as to give voice to moral assurance. This point will need to be developed later.

III

This incomplete catalogue of uses of 'ought' in simple sentences such as 'x ought to do y' has one main point: moral philosophy to date has been insufficiently lexicographical. Even a partial enumeration of the differences already noted between first, second, and third person uses of 'ought' (of which that between a particular first and a particular third person use noted in the discussion of Sartre's example now turns out to be only a particular case) should make us conscious of the need for a far wider range of patterns of analysis than any contemporary writer has so far offered. But, instead of enlarging on this topic here, a possible reply to the arguments that universalizability is not a necessary attribute of moral valuation of the form 'x ought to do y', and that such valuations do not necessarily have a prescriptive function, must be considered. Against these contentions the following counter-argument might be brought.

The essence of moral judgements it might be said is their impersonality. When we judge morally it is at the heart of the matter that we 'do not make exceptions in our own favour' (Kant), that the moral agent must 'depart from his private and particular situation' (Hume). When the moral agent judges an action he judges therefore what anyone should do in that or relevantly similar situations. When he appraises the action of another he thereby commits himself to saying what anyone and *a fortiori* he himself ought to have done. When he decides how to act he thereby commits himself to an appraisal of any similar action by anyone else. Thus appraisal, advice and practical decisions are inexorably linked together. But of these three, practical decisions have the primacy; to appraise someone else's action is to say how he ought to have acted and to give advice is to tell someone else how to act. Moral language, or at least 'ought', is employed *par excellence* in guiding action. In this form the argument brings out the inter-connection of the claim that moral judgements are essentially universalizable and the claim that they are essentially practical and prescriptive. Its force is further brought out by noting a consequence which Hare has drawn from the conclusions of this argument. Hare argues that to say that a man holds a moral principle is to say that he at least sometimes acts on it. A man who claims to believe in keeping promises but habitually breaks them does not in fact hold the principle that one ought to keep promises, according to Hare. Those who have objected to this contention have usually pointed to the problem of ἀκρασία, have argued that if Hare were right we would not have the case of the man whose practice is radically inconsistent with his principles (e.g. P. L. Gardiner, 'On Assenting to a Moral Principle', *P.A.S.*, 1954–5). But this objection takes no cognizance of the way in which the notion of consistency is built into this argument at the theoretical level.

Take the example of a man who appraises actions by one standard and guides his own conduct by another. This differs from the case of the man who is guilty of weakness of will, for such a man's conduct is consistent with his principles, or rather with that set of his principles which he uses to guide his conduct. He merely has two sets of principles. This is sometimes condemned by invoking the maxim 'Practise what you preach' which is also, of course, used to condemn weakness of will. We condemn such a man because and if we disapprove of inconsistency between appraisals and principles of conduct. But while such inconsistency may be morally objectionable, it is not, and the fact that it can be comprehended to such a degree as to be found morally objectionable shows that it is not, unintelligible. Yet on the argument outlined above this is what it must be. For if the meaning of the appraisal 'He ought to have done y' is even partly 'I ought to have done y in those circumstances' (interpreted as 'That is the maxim that would have governed my conduct,' not 'That is the maxim by which I would have appraised my conduct') then the man who asserts that he appraises by one set of principles but acts by another speaks unintelligibly. In other words the view that I am critizing makes consistency between appraisals and principles of conduct a logical requirement. That principles should be so consistent is built into the meaning of moral words such as 'ought'. But the demand for consistency is in fact a moral not a logical requirement. We blame a man for moral inconsistency perhaps, but we do not find what he says meaningless. Appraisals and principles of conduct are logically independent, although in a liberal morality they are required to be morally interdependent. And now we can understand why universalizability is given such a central place by those philosophers whose analyses are directed upon the concepts of liberal morality.

For the requirement that everyone shall be judged by the same standard (the moral counterpart of the political principle that everyone is to count as one and nobody as more than one) in the sense that everyone shall judge everyone else by the standard by which he judges himself is so basic to liberal morality that it is converted from a requirement of morality into a requirement of logic. It is not part of the meaning of 'morality' *tout court* that moral valuations are universalizable, but liberals tend to use the word 'morality' in such a way that this is made part of its meaning. It is worth noting a consequence of this transition from morality to logic, of a kind not unfamiliar in moral argument. If we so characterize moral judgements that we mean by a 'moral judgement' an impersonal one, we make it impossible to approve or disapprove morally of impersonality in judging. For if part of the meaning of 'ought', for example, is such that to say 'x ought' is to say 'x would not be making an exception in his own favour if he . . .' or 'x would be departing from his private and particular situation if he . . .' then to say 'x ought not to make an exception in his own favour' is to utter an empty tautology. This is in essence the same argument as that which Moore used against naturalism, an argument which, as Professor Prior has shown, was anticipated in many ways by Cudworth and Adam Smith. But as Professor Prior has also shown, Moore's argument is not conclusive (*Logic and the Basis of Ethics*, chapter I and *passim*). A tough-minded naturalist can save his position 'by admitting that the assertion that, say, pleasure and nothing but pleasure is good, *is* for him a mere truism; and that if Ethics be the attempt to determine what is in fact good, then the statement that what is pleasant is good is not, strictly speaking, an ethical statement, but only a way of indicating just what study is to go under the name of "Ethics" – the study of what is actually pleasant, without any pretence of maintaining that the pleasure has any "goodness" beyond its pleasantness' (loc. cit., p. 9). Similarly an upholder of the universalizability view of morality could accept the consequence that it is a mere truism to say 'x ought not to make exceptions in his own favour' and contend that all he meant to achieve by this truism was the definition of the field of morality. But if he made this his contention a further consequence would follow (and a similar consequence would follow for Prior's tough-minded naturalist), namely that he would have to abandon any claim to be offering us a neutral logical analysis of moral language. For plainly ordinary moral agents do disapprove of making exceptions in one's own favour in non-truistic fashion (just as they hold that pleasure is good in similar fashion). To assert that universalizability is of the essence of moral valuation is not to tell us what 'morality' means or how moral words are used. It is to prescribe a meaning for 'morality' and other moral words and implicitly it is to prescribe a morality.

Finally, one more feature of the prescriptive theory of moral valuation must be examined briefly. A maxim may be said to prescribe or to guide conduct in one of two ways. Clearly we might describe conduct that accorded with a maxim as guided by it and speak of it as the conduct prescribed by the maxim, if we were willing to adduce the maxim to justify the conduct. A man who habitually kept his promises might when challenged on a particular occasion as to why he had put himself out to meet a friend or to pay a debt avow 'One ought to keep one's promises.' It would be natural to describe the man's conduct as guided by the maxim. But this is not to say that up to the point where justification was demanded the maxim ever entered his thoughts, at least since the time that he learnt it as a child. And thus the maxim guides conduct in a sense quite different from that in which a maxim may be said to guide conduct if we explicitly consult it when perplexed as to what we ought to do. Most of the actions discussed in moral philosophy textbooks – promise-keeping, truth-telling and the like – are in practice carried out without any sort of conscious reference to maxims. So that in the more explicit sense of 'guide' where part of what we mean by 'guide' is 'to give guidance', 'to tell us what to do', the relevant maxims do not guide us when we keep promises, or tell the truth. They do not guide us because we do not need to be guided. We know what to do. We tell the truth and keep promises most of the time because it does not occur to us to do otherwise. When we are tempted not to do these things from some motive of self-indulgence, it would still not be true to say that if we resist the temptation, the maxim guided our conduct. What guided our conduct was our decision to abide by the conduct prescribed by the maxim. So that in this sense 'x ought to do y' prescribes a certain line of conduct but it does not guide us or tell us what to do. That this is so is even more obvious when we consider those cases where we have already noted a common use of 'ought', cases of moral perplexity. When Huckleberry Finn wrestles with the problem of whether to return Jim, Miss Watson's slave, he is not guided by the maxims of his morality for his whole problem is whether to abide by those maxims or not. The maxims tell him that property is sacred, and that Jim is merely property. Nor is Huck guided by a new set of maxims, perhaps the anti-slavery maxim that 'One ought to treat no human being as a slave'. For Huck retains the same general attitudes as he was brought up with. (When someone inquires if anyone has been injured in a steamboat accident, he says 'No'm, killed a nigger.') In fact in deciding not to inform on Jim he feels wicked and thinks of himself as wicked. He thinks of himself in fact as making an exception in his own favour, of favouring his own friend at the expense of morality. He finds his way morally by means of an only half-articulate sympathy. But he does not find it by universalizable maxims or indeed by maxims at all.

When you leave the ground of conventional morality, you leave the guidance of maxims behind. Yet it is just here that one needs guidance. Where men pass from one set of maxims to another, or act morally without maxims, there is an area where the logician and the linguistic analyst are necessarily helpless. For they are not presented with the kind of material which they need for analysis. Only the phenomenologist can help us here and the kind of phenomenology we need is that supplied by the novelist. It is because the moral philosophers of existentialism have been primarily concerned with this kind of situation that they have so often resorted to the novel. For all that can be done is to exhibit the passage of the moral agent through perplexity. To offer us a maxim on which or in accordance with which the moral agent finally acted is to tell us what the resolution of perplexity was but not how the perplexity was resolved. In this clear-cut sense then the maxims of morality do not guide us, nor do they prescribe conduct to us. And to describe the function of moral valuation in general or of 'ought' in particular as prescriptive is highly misleading unless this is made clear. The catalogue of possible uses of 'ought' needs to be supplemented by a catalogue of those moral purposes for which 'ought' and words like it and sometimes any words at all can be of little or no use.

One last point relating moral perplexity to the claims made for moral judgement as prescriptive and universalizable: here, as everywhere in moral philosophy, much depends on the choice of examples. Where there is real moral perplexity it is often in a highly complex situation, and sometimes a situation so complex that the question 'What ought I to do?' can only be translated trivially into 'What ought someone like me to do in this kind of situation?' This is important because this translation is often not trivial at all. When I am puzzled it is often useful to pick out the morally relevant features of the situation and of my position in it and, having isolated them from the particular situation, I am in a better position to solve my problem. But, where a situation is too complex, phrases like 'someone like me' or 'this kind of situation' become vacuous. For I am the only person sufficiently 'like me' to be morally relevant and no situation could be sufficiently like 'this kind of situation' without being precisely this situation. But what situation could be complex in this way? The situation of Françoise in Simone de Beauvoir's *L'Invitée* or that of Mathieu in Sartre's *Les Chemins de la Liberté* are examples that spring to mind, for part of their problem is to decide which features of their situation are relevant; part of their problem is to discover precisely what their problem is. And this brings out the point that it is because Sartre and Simone de Beauvoir are concerned with morality of this kind and in this way that they present and can only present their insights in the form of novels rather than of logical analyses.

3

MacIntyre on defining morality

W. K. FRANKENA

In 'What Morality is Not', *Philosophy*, XXXII (1957),[1] Mr Alasdair MacIntyre argues against the view, now common, 'that universalizability is of the essence of moral valuation'. On page 331[2] he uses an argument which is an adaptation and extension of Moore's naturalistic fallacy argument, and which is generalizable. As Moore's argument, if cogent, holds against all definitions of 'good', 'right', etc., so MacIntyre's argument, if good, holds against all definitions of 'moral' and 'morality'. For this reason I shall examine his argument, as I once examined Moore's.[3] I wish to do this partly because I should like to go on looking for definitions or elucidations of words like 'moral' such as are contained in the universalizability view, and partly in order to air some general questions which are raised by what MacIntyre says.

MacIntyre's argument comes in two parts. The first is contained in the following passage:

> If we so characterize moral judgements that we mean by a 'moral judgement' an impersonal one, we make it impossible to approve or disapprove morally of impersonality in judging. For if part of the meaning of 'ought', for example, is such that to say 'X ought' is to say 'X would not be making an exception in his own favour if he . . .', . . . then to say 'X ought not to make an exception in his own favour' is to utter an empty tautology.

[1] Reprinted on p. 26 of this volume. *Eds.* [2] p. 36 of this volume. *Eds.*

[3] See 'The Naturalistic Fallacy', *Mind*, 1939 (reprinted in P. R. Foot (ed.), *Theories of Ethics*, Oxford, 1967). Moore's argument is to be found in *Principia Ethica* (Cambridge, 1903). *Eds.*

Reprinted from Philosophy, *1958, pp. 158–62, by permission of the author and* Philosophy.

Here the second statement is given as a reason for the first. This means that MacIntyre is supposing that if I hold that 'J is a moral judgement' means in part 'J is an impersonal judgement', then I am also holding that part at least of what J says is that it, J, is an impersonal judgement. Now this is a mistake, though perhaps a natural one. Suppose I hold that '"X ought to do A" is a moral judgement' means in part '"X ought to do A" is an impersonal judgement'. It does not follow that part of the meaning of 'X ought to do A' is 'X would be proceeding impersonally if he were to judge that he should do A (or were to do A)'. For even if part of what '"X ought to do A" is a moral judgement' says is that 'X ought to do A' is an impersonal judgement, it does not follow that this is part of what 'X ought to do A', taken by itself, says.

MacIntyre, like other contemporary writers, is making the mistake of thinking that to define 'moral' is also to define 'ought'. This is natural, since when we speak of moral action (as versus immoral action) we mean action which is right or obligatory. But what is in question is the meaning of 'moral' as applied to judgements, and here 'moral' is not equivalent to 'right' or 'obligatory'. To say that 'X ought to do A' is a moral judgement is not to say or even imply that part of what it says or means is that it is a moral judgement. To say that '"X ought to do A" is a moral judgement' means or entails '"X ought to do A" is an impersonal judgement' is only to give an analysis of 'moral' as applied to judgements; it is not to give an analysis of 'ought' as it appears *within* such judgements. An elucidation of the logic of 'moral' here is not necessarily, even in part, an elucidation of that of 'ought'.

But then the first part of MacIntyre's argument collapses. The point made in his second sentence does not support the claim made in the first, since it rests on a mistake. But let us look at the first sentence by itself. It says, 'If . . ., then . . .' Is this allegation true? One might hold that it is, even independently of the consideration advanced in the second sentence. Again, so far as I can see, one would be involved in a mistake if he did. Suppose we say that a moral judgement is one which is made from the moral point of view, and that this point of view, among other things, is an impersonal one. Does it follow that one is uttering a tautology if he says, 'We ought (morally) to take this impersonal point of view'? Not at all. For to speak from an impersonal point of view is not to *state*, even in part, that one is doing so; at most it is to *express* an impersonal attitude or orientation. One may not be stating anything, as some emotivists have held in the case of moral judgements, or one may be stating something very different, as one is when one makes a scientific assertion. (Here one is taking an impersonal point of view, perhaps even contextually implying that he is doing so, but not stating or logically implying that he is.) But if this is true, then we may morally approve of impersonality in judging without uttering a tautology. What we are doing then is this. We take the impersonal point of view ourselves and from this point of view we judge that others and we ourselves on other occasions ought to be impersonal in our judgements. This is perfectly possible both logically and psychologically on the theory in question. And in the interests of morality it is not pointless, and need not be felt to be so, to pass such judgements on occasion.

We come now to the second part of MacIntyre's argument. He recognizes that a tough-minded naturalist can answer Moore's argument against his definition of 'good' in terms of (say) 'pleasure' by saying that 'Pleasure is good' is for him a mere truism marking off the field of ethics. 'Similarly,' he says:

> an upholder of the universalizability view of morality could accept the consequence that it is a mere truism to say 'X ought not to make exceptions in his own favour' and contend that all he meant to achieve by this truism was the definition of the field of morality.

Then he states the second part of his argument, with the suggestion that Moore could complete his case against naturalism in the same way.

> But if he made this his contention a further consequence would follow . . . namely that he would have to abandon any claim to be offering us a neutral logical analysis of moral language. For plainly ordinary moral agents do disapprove of making exceptions in one's own favour in non-truistic fashion. . . . To assert that universalizability is of the essence of moral valuation is not to tell us what 'morality'

means or how moral words are used. It is to prescribe a meaning for morality and other moral words and implicitly it is to prescribe a morality.

In so far as this part of the argument rests on the claim that ordinary moral agents do disapprove of making exceptions in one's own favour, we have already dealt with it. Even if this is a fact, we have seen how it can be explained on the view that a moral judgement is an impersonal one. But some other points are raised by this second passage. (1) Besides depending on the fact or alleged fact just mentioned, MacIntyre appears to assume that the view that a requirement to be willing to universalize is one of the rules of our actual moral language *cannot* be true. But for him simply to assume this is to beg the question, for his opponents are or may be holding precisely that this is true. If any neutral statements describing the rules of our use of terms like 'moral' can be correctly made, why not this one? In the rest of his essay MacIntyre gives independent reasons for thinking it to be incorrect, but here he seems to suppose in advance that it must be so. This is simply a *petitio*, unless it is being maintained as part of a general view that no neutral logical analyses of moral language are possible.

(2) MacIntyre does seem to be taking this more general position. I shall not ask whether this is itself a neutral position or not, though presumably one might. It must be remarked, however, that we are not now asking about the analysis of first-order ethical words like 'right' and 'ought', but about the analysis of second-order words like 'moral' (as applied to judgements) and 'morality'. Even if no neutral analyses of the former are possible (and nothing in MacIntyre's argument serves to prove this, since his thinking on this point is exactly parallel to the argument we are discussing and finding flaws in), it does not follow that no such analyses of the latter are possible – unless we presuppose the yet more general view that no neutral analyses whatsoever are possible. If any neutral analyses are possible, I see no reason why they should not be possible in the case of 'moral' in such phrases as 'moral judgement', 'moral rules', 'moral institutions', etc.

(3) It is surprising to find a member of the movement to which one takes MacIntyre to belong apparently maintaining that no

neutral statements can be made about our use of moral language. But in the second passage he seems also to subscribe to a position which is more common among members of that movement, namely that philosophers must deal in neutral logical analyses, and that one can dispose of a suggested analysis by showing it is not neutral. For if this is not so one can accept what he says here and go merrily ahead to define moral judgements as universalizable, etc. (provided, of course, that one can answer his other arguments against such a definition). Now this conjunction of positions is interesting. First, many people, on becoming convinced that no neutral analyses of moral terms are possible, would conclude that then moral philosophers may or must go ahead to give non-neutral ones, defending them by any means that may be found appropriate – a conclusion which seems entirely reasonable. Second, from the two positions conjoined it follows that philosophers must give up looking for any analyses or elucidations of our use of moral language. Does MacIntyre really mean to hold this? His only alternative would seem to be to hold that 'morality' stands for an indefinable property, as Moore held that 'good' does. However, if he maintains the former alternative, what is he doing in the rest of his essay? If he is there offering neutral elucidations of our moral language, why cannot his opponents make the same claim and possibly be right? If he does not mean to take the first alternative, which of the two conjoined positions will he give up?

(4) Are non-neutral 'analyses' philosophically wicked, as MacIntyre and others suppose? When is an analysis non-neutral? Presumably, when it is not simply an explication of the rules according to which we use the bit of language in question. Now, as I said, if such mere explications of the actual rules of our usage are possible at all, I see no reason why they should not be possible in the case of 'moral' or why the universalizability rule cannot be one of them. But suppose that the analyst not only describes a rule of actual use but also subscribes to this rule and prescribes it to others, as analysts very often seem to do, wittingly or not. Then, I take it, his analysis is no longer neutral. But is it philosophically wicked? If so, why? If not, why cannot those who define 'morality' argue that they are non-neutral only in this sense? There is the further question whether or not it is a

philosophically unpardonable sin to propose a revision of the rules for our use of moral language. I am not convinced that it is, but shall not argue this now.

(5) MacIntyre takes for granted also, in the last two sentences quoted, that if a putative analysis of 'moral judgement' is non-neutral or prescriptive, then it must be a disguised *moral* judgement. So have many of Toulmin's critics. But is this necessarily the case? Is saying that 'moral judgement' means 'impersonal judgement' necessarily a moral judgement even if it be granted that it is a normative one? Not all normative or prescriptive judgements are moral judgements. Again, not all definitions are disguised moral judgements, even if they can be defended only on pragmatic grounds. Besides moral imperatives, there are, as Kant said, technical and prudential ones. Now, so far as I can see, when we define terms like 'science', 'scientific judgement', etc., we are making neither a disguised scientific judgement nor a disguised moral one. We may be uttering a disguised imperative, but it is technical or prudential, and not moral. Why is the case in principle different when we define terms like 'morality' and 'moral judgement'? If we are not debating a substantive scientific question in the first case why can we not claim that we are not debating a substantive moral question in the second? Defining terms like 'moral judgement' may be part of an attempt to understand, re-think, and possibly even to revise the whole institution which we call morality, just as defining 'scientific judgement' may be part of an attempt to do this for science. Must we regard the former enquiry as moral in the same sense in which 'Should I keep my promise?' is moral, when we do not regard the latter enquiry as scientific in the same sense in which 'Does water freeze at 30° Fahrenheit?' is scientific?

It is true that some people are for impersonality of judgement in morals and others against, and that this debate comes under the heading of ethics. But the issue between them need not be interpreted as a strictly moral one. One side may be saying that impersonality is part of what is to be meant by calling a judgement moral, and the other that it is not. Then their quarrel is not moral in the sense of being *within* morality; it is only moral in the sense of being *about* morality. They cannot argue it as if it were an issue which could be settled by an appeal to principles provided by morality; they must go back to look at the whole institution of moral discourse and see what definitions of it seem to accord best with the purposes for which it exists. The argument is not one which is being carried on within the framework of an institution which is being left unquestioned; it is one which is carried on at a level which asks what character this very institution is to have and what place it is to take in our economy of institutions.

In attacking this argument of MacIntyre's I do not mean to disparage the rest of this interesting essay, in which he makes a good case for a more existentialist view of moral judgements than that which prevails among British and American moral philosophers. I simply could not permit this latest use of the 'naturalistic fallacy' to go unchallenged, any more than I could the first one.

4

Critical notice of R. M. Hare's Freedom and Reason

C. C. W. TAYLOR

The theory of moral reasoning which Mr Hare expounds in this book is presented as an attempt to resolve a dilemma or antinomy which, as he maintains in his opening paragraphs, confronts anyone who is faced with a serious moral problem. The dilemma is constituted by two necessary features of such a situation, on the one hand 'that a man who is faced with such a problem knows that it is his own problem, and that nobody can answer it for him' (p. 1), and on the other 'that the answering of moral questions is, or ought to be, a rational activity' (p. 2). Now it is undeniable that these features do characterize the situation of a moral agent deliberating on action; a rational moral agent is one who solves *his* problems on rational grounds. What is not clear is why these features should be thought to constitute an antinomy. For, taken at its face value, the first seems merely to point out that every agent is an individual agent, with the consequence that, given an agent X in circumstances which call for some action of his, whatever course he takes will be describable as an action of X. Yet, obviously, this feature of being an agent is in no way incompatible with acting rationally. It appears, then, that if Hare's dilemma is to be a genuine one the first horn must be interpreted more widely than its actual wording would suggest. Other remarks in these opening paragraphs suggest that Hare has the following

---

Reprinted from Mind, *1965, pp. 280–98, by permission of the author and* Mind.

interpretation in mind, that the question what one ought to do in a given situation can never be conclusively settled by appeal to the facts of the situation, since it is always open to the agent to decide without logical error that any set of facts does or does not constitute sufficient grounds for action. But far from representing, as Hare claims, a 'conviction which every adult has' (p. 2) this view appears quite opposed to the moral consciousness of the ordinary man, who is apt to take it for granted that from the fact that an action has such and such characteristics it follows that it is right or wrong, and requires quite a lot of philosophizing to argue him out of this belief. Hare appears to admit this when he says (p. 2) that in order to get the dilemma going at all one must assume that there can be no logical deduction of moral judgements from statements of fact. That is to say, this dilemma appears to arise only if one commits oneself to a particular philosophical position about the nature of moral judgements. But the suggestion of these opening pages is that it is necessary to adopt Hare's theory of moral reasoning in order to resolve a dilemma which forces itself upon anyone who does any serious moral thinking. Obviously, the fact that the dilemma does not arise in this way does nothing to invalidate the argumentation of the book as a whole. Yet Hare's presentation of the dilemma is to this extent misleading, that it represents those who reject the first horn as thereby denying moral freedom, a charge which seems quite unwarranted. To take an example from a related form of reasoning, in order that the jury should be free to make up their own minds about the guilt of the accused, it does not have to be the case that there is no possibility of his guilt being conclusively established by the facts. Hare's dilemma does not in fact arise from the concept of moral freedom at all, but from the quite distinct concept of moral autonomy. His problem is to show how, given that no moral judgement can follow from any set of factual propositions, the making of such judgements can nevertheless be a rational activity. To the solution of this problem the rest of the book is devoted.

The solution is in terms of the theory of moral judgements as a class of universalizable prescriptive judgements which was introduced in *The Language of Morals*. The form of its 'presentation' here is as follows. Chapters 2–5 of part I, give a general recapitulation of the fundamentals of the theory, with some refinements and answers to objections. Part II gives a general account of its application to moral argument, showing the force of arguments deriving from the theory in providing a rational basis for the reconciliation of interests and considering their weight in questions of conflicts of ideals. Finally part III shows the theory applied to a particular moral problem, that of the oppression of people on grounds of racial difference. In what follows I shall be concerned mainly with parts II and III; this is not to imply that part I is without interest, but merely that, covering as it does fairly familiar ground and being in addition presented with Hare's characteristic and admirable lucidity, it is in the main unproblematical. The later parts, while not less lucid, both break some new ground and raise some questions of crucial importance for the theory as a whole.

Some particular points from part I are nevertheless especially worth noting. Thus in his discussion of universalizability in chapter 3 Hare refutes objections to this thesis made on the grounds that, every action being unique, it is impossible to pass from a judgement about a particular action to a general principle about actions done in similar circumstances. The difficulty of describing adequately a particular action, he points out, is that of finding a description specific enough to include all the morally relevant features; there is, however, no incompatibility between a description's being specific and its being universal in the sense of including only general terms without proper names and particular spatio-temporal indicators. As Hare says (p. 39), 'The opposite of "general" is "specific"; the opposite of "universal" is "singular"'; objections to universalizability on the above grounds appear to have arisen from a confusion of these two distinct contrasts.

Again, the few pages (61–6) on free will, though merely a parenthesis in a discussion of the conditions in which it is logically proper to consider the question of whether one ought to do some action, are among the most valuable contributions which have appeared on this topic. Hare shows that, given that we find ourselves and others in the situation of having to decide what to do in a practical context, we have logically no alternative to treating ourselves and them as responsible agents in that context. No degree of success in predicting, on whatever grounds, the outcome of an agent's decision can free him from the burden of actually deciding, since even if he is aware of the prediction he has to decide to do what the prediction says he will do. Of course, this account holds only where a decision is operative in bringing about the action; if we had a theory which enabled us to predict that no matter what the agent decided, including the case of his making no decision at all, he would still act in a particular way, then there would be no ground for insisting on any measure of freedom. The effect of Hare's discussion is to show that nothing less than this will suffice for a genuine determinism, and to highlight the implications of such a theory, in particular the revision which it would require of accepted views of moral education.

In the context of Hare's theory of moral judgements the most crucial section of part I is chapter 5, where he discusses the difficulty raised for prescriptivism by the case of someone who, while claiming sincerely to assent to a moral principle, nevertheless fails to act on it in a situation where there is no apparent obstacle to his doing so except his own moral weakness. This is clearly a crucial difficulty for Hare, for since the criterion of a word's having prescriptive meaning is that anyone who uses the word with that meaning commits himself to assenting to some imperative entailed by the proposition containing the prescriptive word, and the criterion of assenting to an imperative is acting on it, any case of the use of a word with prescriptive meaning which does not result in action in the appropriate circumstances provides a counter-example which falsifies the theory. Hare sees this position clearly, and attempts in this chapter to answer the difficulty; I confess, however, that I do not follow his argument at the beginning of the chapter that the recognition of the existence of a problem of moral weakness counts as *prima facie* evidence *in favour of* prescriptivism. On page 68 he argues as follows: 'if moral judgements were not prescriptive, there would be no problem about moral weakness; but there is a problem; therefore they are prescriptive'. On the face of it this is parallel to a naive attempt to dissolve the theological problem of evil which runs:

'If God were not infinitely loving there would be no problem of evil; but there is a problem of evil; therefore God is infinitely loving.' Yet clearly something more significant lies behind this, namely a recognition of the fact that anyone who agrees that the function of moral language is to guide conduct is faced with the problem of accounting for apparently sincere uses of such language which yet fail to fulfil their conduct-guiding role. It seems, however, quite unwarranted to cite this fact in favour of the prescriptivist theory of how moral language functions to guide conduct; in order to justify that inference we should require evidence that no other theory is capable of accounting adequately for that function. But in that case the existence of a problem of weakness of will does nothing new to convince us of the necessity of adopting prescriptivism, since we had adequate grounds already for concluding that no other theory would do what was required of it.

Hare's answer to the prescriptivist's dilemma is simple and bold. In every case of weakness of will the agent is literally unable to carry out the moral principle to which he verbally assents, being prevented from doing so by a psychological force which is properly to be described on the model of physical compulsion. The evidence offered in support of this assertion is, firstly, that the Greek *akrasia* and the English 'weakness of will' both imply inability to carry out one's moral principles (if one is too weak to do something then one is unable to do it), and, secondly, that two famous descriptions of this situation, Ovid, *Met.* vii, 9–21 and *Romans*, vii, 14–25, both imply that the agent is unable to act on the moral principle to which he assents. But if this question is to be settled by appeal to what people ordinarily say, it is obviously in point to bring against these examples the common conviction that in such a case it was not beyond one's power to act according to one's principle, since one could have done so had one chosen to, and could moreover have chosen to had one wanted to. Unless, then, Hare is prepared to rest his case on appeal to the authority of St Paul and Ovid (an oddly-assorted pair, it might be thought), together with a somewhat dubious point of idiom (for it has yet to be shown that the theoretical picture underlying the use of the terms 'weakness of will' and *akrasia* is a truer picture than that which underlies the conviction that one could have acted on one's principle), he must show that the description which the common conviction embodies is mistaken, and that the situation is more properly to be described in terms of inability to act by reason of psychological compulsion. This he does not attempt to do. It must therefore be concluded either that he has simply overlooked the necessity of disposing of the common conviction or that he does not let it count against his thesis on the ground that, whatever the evidence, it *must* be the case that the weak-willed man is unable to act on his principle. In the latter case the thesis clearly represents, not a solution of the problem, but a resolution to ignore it. In the former, the thesis, while not indeed empty, is left virtually unsupported.

An argument which might be brought against the common conviction is the following. The true description of a case of weakness of will requires the admission that, while X could indeed have acted on his principle had he chosen to do so, and could have chosen to do so had he wanted to, nevertheless he was unable to want to, and so unable to choose to act on his principle, and so unable to act on it. We are thus able to give a sense to Hare's notion of psychological force analogous to physical restraint preventing the weak-willed man from acting on his principle; this force is whatever force it is which makes him incapable of wanting to act on the principle more than he wants to contravene it. But how do we know that there is such a force? What justifies the insistence that in this case the agent *was unable to want* to act on his principle, rather than that he *did not want* to act on it? If this insistence is to have any content, much less be proved true, we must be able to attach a sense to the idea of being restrained or prevented from wanting something. Now restraint gets its sense from the situation in which an agent is prevented from doing something which he wants or has decided to do; but if we attempt to apply this model to being prevented from wanting to do something we reach the odd conclusion that the agent wants or decides to want to do x and is then prevented from putting this want or decision into operation. But in so far as one can give a sense at all to the idea of wanting to want to do x, this seems just a particular sort of wanting to do x. Thus if I have an ambition to climb Mount Everest, together with a strong revulsion at the thought of the privations involved, I might school myself to overcome this revulsion and instead to take pleasure in the thought of actually struggling to the top. While this process was going on I might indeed describe myself as wanting to want to climb Everest, but this would have to be cashed as (putting it crudely) having the wish or the intention to have the urge or the desire to climb Everest. But I could not have that wish or intention unless I had the wish or intention *to climb Everest*; i.e. wanting to want to do x entails (in some sense of 'want') wanting to do x. This sense of wanting to want to do x still seems insufficient to give a sense of being prevented or restrained from wanting to do x adequate for Hare's view of *akrasia*; for we have now to distinguish the case of the man whose intention to desire to do x is prevented from being put into effect from that of the man who does not intend as strongly as he could have done to desire to do x. That is to say, the same problem has arisen one stage further back. It will not do simply to posit a psychological force to distinguish the cases; for if we have no criterion for when the force is operative we might as well say that the force just is the state of not intending as strongly as one could have done to want to do x. Thus it appears that the introduction of the notion of being unable to want to act on one's principle, in the sense of being subject to some psychological force which prevents one from acting on it, is insufficient to show that the popular view of *akrasia* as failing to act on one's principle when one could have done so is mistaken. Hare's dismissal of this view remains, then, without support.

The main defect in Hare's treatment of *akrasia* is that the key concept of 'psychological compulsion' is left undefined, and in particular that no criteria are suggested for differentiating it from simply not wanting to act on one's principle more than one wants to contravene it. It would obviously be possible to remedy this defect in the terms of some psychological or (more plausibly) neuro-physiological theory; thus being unable to act on one's principle might be defined in some such way as this, that given a certain state of the organism, e.g. a state of the brain, it could be predicted with certainty according to recognized physical or other laws that the organism would not produce the stimuli necessary to bring about action in accordance with the principle. It seems *prima facie* likely that a great many cases of *akrasia* are of this sort; there is a strong inclination to say that the alcoholic struggling with his desire for a drink is incapable of resisting the temptation. But at the same level of ordinary talk, there is an equally strong presumption against applying that analysis to the man who, while holding that one ought not to walk across the college lawn, does so occasionally when he is in a hurry; and cases of this latter sort are just as much counter-examples to the prescriptivist thesis as the more spectacular sorts of delinquency which are generally discussed in this context. But even granted that such cases could plausibly be brought under such a theory, we should require (*a*) criteria for identifying the particular states which are constitutive of psychological compulsion and (*b*) reasons for concluding the theory to be true. Condition (*a*) implies the further condition that the theory should not be one which purports to show that in every case of action the agent was in some sense incapable of doing anything other than what he in fact did. For such a theory would once again eliminate the contrast between not wanting to act and being unable to act on which depends the prescriptivist solution of the problem. There appear no grounds for asserting in advance that a theory satisfying these conditions could not be devised; some thesis of the type of Hare's might then turn out to be true for some or even all cases of *akrasia*. But to base an answer to the problem on the thesis that we know already that some such theory is true would be to do a quite unjustified piece of *a priori* psychologizing. I conclude, then, (*a*) that Hare himself provides no adequate answer to the objection presented to prescriptivism by the phenomenon of *akrasia*, and (*b*) that the only plausible answer to this objection has at best the status of an unsupported conjecture, since it requires factual support which is not available in the present state of scientific knowledge. Clearly, these considerations do not make it impossible to hold on to prescriptivism; they do, however, suggest that it is an urgent task for the prescriptivists to produce real evidence of psychological compulsion in such cases as that of the man who crosses the lawn while thinking that he ought not to, and further that it would be more prudent to look for a theory which does not commit science in advance. Just how much weight one gives these considerations depends on how firmly one is convinced of the value of prescriptivism in general; their persuasive force will clearly be increased if weaknesses appear in the theory independently of the problem of *akrasia*.

That certain weaknesses are indeed to be found appears from the schema of moral argument which is proposed in part II, chapter 6, sections 3 ff., and developed in the rest of part II and in part III. The validity of this argument is central to Hare's whole thesis, since he is rightly insistent that the test of the value of an ethical theory is its utility in leading to the solution of actual moral problems. The argument runs as follows. From the prescriptivity of moral language it follows that anyone who gives his moral approval to an action of a certain sort commits himself to the imperative 'Let action A be done in the appropriate circumstances.' It further follows from the universalizability of ethical judgements that in committing himself to that imperative he also commits himself to the imperative 'Let me, in the appropriate circumstances, be the victim of action A.' And in cases where action A involves the infraction of the sort of interest which most people regard very highly it is unlikely that anyone but an extreme fanatic would be prepared to assent to this prescription. We thus have a form of argument which is sufficient to make it clear beyond doubt to most people that the infringement of other people's major interests is morally wrong, which is in itself a significant achievement to the credit of moral philosophy.

In the majority of cases this argument will do all that Hare claims for it. There is, however, a class of cases to which it will not apply, a class which, unluckily enough for Hare, is typified by the very sort of case which he has chosen to demonstrate the effectiveness of his argument, viz. racial discrimination. It therefore seems that we should look for some reformulation of the argument which will remove this restriction on its generality. The restriction arises from the following feature of the argument, that it is necessary for Hare that his opponent should be obliged to assent *in propria persona* to an imperative prescribing that a certain action be done to *him*, given that he is in the same position as the victim of the action at present in question. In certain cases, however, it is logically impossible for Hare's opponent to be in the victim's position, since anyone who had the characteristics of the victim would be a different person from the man who is now Hare's opponent. And since there is no clear reason why he should have overwhelming objections to prescribing that the interests of someone else should be infringed, which is indeed precisely what he is prescribing in the present case, the argument loses all force.

The reason why it is necessary for Hare's argument that his opponent should commit himself *in propria persona* to the command that something unpleasant should be done to *him* is that Hare counts as a refutation in a practical dispute nothing less than that one of the parties should be shown to have committed himself to incompatible imperatives. This is clearly brought out by the following passage from page 108:

> First of all, we shall make the nature of the argument clearer if, when we are asking *B* to imagine himself in the position of his victim, we phrase our question, never in the form 'What *would* you say, or feel, or think, or how *would* you like it, if you were he?', but always in the form 'What *do* you say (*in propria persona*) about a hypothetical case in which you are in the victim's position?' The importance of this way of phrasing the question is that, if the question were put in the first way, *B* might reply 'Well, of course, if anybody did this to me I should resent it very much and make all sorts of adverse moral judgements about the act; but this has absolutely no bearing on the validity of the moral opinion which I am now expressing.' To involve him in contradiction, we have to show that he *now* holds an opinion about the hypothetical case which is inconsistent with his opinion about the actual case.

The strength of this argument is to make it clear that the only way of convincing someone on a moral issue is to show him that he actually *has* overriding reasons for acting in a certain way, not merely that he would have such reasons were he in some different situation. Its weakness lies in its assumption that the only way of showing someone that he actually has such reasons is by making him actually pronounce upon a hypothetical situation which may be logically precluded from ever occurring.

To see how it may be logically impossible for Hare's opponent to be in the position of his victim, we should consider his own case of racial discrimination. How would his argument deal with a white man who said 'Africans who have been brought up in tribal society, whose moral outlook is consequently quite different from ours, who don't accept European standards of culture, education, etc., are not worth the slightest moral consideration, but should be treated simply as chattels by their white masters'? There is no point in saying to this man, suppose him to be a South African farmer of Dutch descent, moderate means, secondary school education and rigid Calvinist principles, 'What do you, Henrik Potgieter, say of a hypothetical situation in which *you*, Henrik Potgieter, are in the situation of such an African?' For what can count as being in the situation of such an African other than having not only the physical characteristics but also the upbringing, outlook, sympathies and interests of such an African? And what is that other than actually being such an African? But what is the sense of the supposition that a white farmer of Dutch descent, etc., might in certain circumstances be an African of a totally different educational level and moral and social outlook? Surely the only sense is that the farmer has come to be another person. But the whole force of the argument depended on the assumption that the person in the hypothetical situation and the person required to legislate for that situation should be *the same* person, in order that that person should find himself obliged to prescribe *for himself* something which he does not want. Given, then, that the victim in the hypothetical situation logically cannot be himself, all that he is being required to do is to prescribe that some other, hypothetical person should be victimized, and since he has no scruples about victimizing actual persons it is hard to see why he should be expected to feel more strongly about a merely hypothetical person.

To meet this objection it is necessary to show that there is no logical impossibility in the agent's being in the position of the victim of his action. Yet it is not clear how such an assertion is to be supported; clearly a piece of science fiction such as that suggested at page 218 will not suffice, since what is in question here is not merely someone's acquiring a new physical characteristic but his having a different heredity and upbringing from the one which he in fact did have. Nor is it profitable to make the move suggested at page 171, that of pretending to be in possession of evidence that, all appearances to the contrary, our white farmer is in fact an African; for what is the sense in the supposition that someone who has none of the characteristics which differentiate Africans from whites is none the less an African? What seems to be needed is a theory of personal identity which could allow that two men with completely different life histories and with distinct physical and psychological characteristics might yet be the same person. This, however, seems an unacceptable supposition, since in this case all the criteria for personal identity have been stripped away, leaving only a sort of sameness which is sameness in no respect, *i.e.* no sameness at all.

Against this it may be urged that if this argument were correct it would be impossible to imagine that one were someone other than the person one is; but, as ordinary experience shows, nothing is easier to imagine than this; therefore the argument is fallacious. This counter-argument, however, depends on a confusion between three different imaginative states: (*a*) imagining that one is some person other than the person one in fact is, (*b*) 'putting oneself in someone else's place', i.e. imagining what one would oneself feel were one in the same situation as someone else is in, or in a similar situation, and (*c*) imagining what someone else feels in a particular situation. Of these states, (*a*) is logically impossible for anyone not suffering from delusions about his identity; thus if I, being someone other than Napoleon, imagine that I am Napoleon, then that fact is itself a sufficient criterion of my not knowing, or perhaps having temporarily forgotten, who I am. There is of course a sense in which I can imagine that I am Napoleon without believing that I am Napoleon, as (to borrow an example of Hare's own) officers studying the Waterloo campaign might be asked to imagine that they are Napoleon, but that seems simply to mean trying to look at the situation as it must have appeared to Napoleon, which is included in state (*c*) above. In general, anyone who claimed that, while retaining knowledge of his true identity, he was imagining a situation in which he was identical with some person other than the person he in fact is, must either be misdescribing states (*b*) or (*c*) above or be talking nonsense. For the word 'he' in 'Smith is imagining a situation in which he is Napoleon' must refer to some person or other. If it refers to Napoleon, then Smith is imagining a situation in which Napoleon is Napoleon, which appears to be a roundabout way of saying that Smith is imagining some situation involving Napoleon, i.e. state (*c*). If it refers to Smith, then Smith is imagining a situation in which Smith is Napoleon; but since there is nothing which could count as that situation's obtaining, the situation cannot be imagined either, since to imagine it is to imagine it *as obtaining*. State (*c*), on the other hand, is never logically impossible, though it may be practically very difficult or even impossible. This state is, furthermore, ethically the most important, since it enables one to cultivate towards the widest possible range of cases the kind of imaginative sympathy which leads one to act well towards other persons. State (*b*) is more restricted in its range, for there are some situations such that if one were to be in them one would have to be someone other than the person one in fact is, and hence imagining oneself in that situation would involve being in state (*a*), which we saw to be logically impossible for someone not suffering from delusions. One might be tempted to think that cases where state (*b*) is logically impossible cannot be cases of genuine ethical importance, but this is far from the truth. Cases of sex-change notwithstanding, it is logically impossible that a middle-aged man should be able to imagine what *he* would feel were he a girl of seventeen receiving a letter from her lover telling her that he is leaving her, but given that he is sympathetic enough he may be able to imagine what she will feel, perhaps partly through recalling his own feelings at seventeen. It may, furthermore, be extremely important that he should be able to imagine this, for he may be the girl's father, or for that matter the lover himself, and it may be necessary that he should be able to imagine her suffering if he is to regard it as something which *he* wants to avoid, and hence as a reason for him to act in a certain way.

These distinctions might be regarded as trivial, on the ground that, though Hare may so have phrased his original argument as to suggest that state (*a*) is logically possible for someone not deluded as to his identity, his argument could be rephrased in terms of states (*b*) and (*c*) without the slightest embarrassment. His hypothetical opponent is then being challenged to imagine how his proposed victim will feel if he acts as he proposes, if possible by imagining how he would feel in a similar situation; the point of issuing this challenge is to induce him to regard the suffering of his victim as a reason for him not to do the act which will cause the suffering. To take this line is, however, to overlook Hare's insistence that his opponent must be made to pronounce *in propria persona* an imperative with respect to a hypothetical situation in which he is himself the victim; the whole point of this insistence is to ensure that the person issuing the imperative and the victim of the action enjoined by the imperative are the same person, since only thus can we conclude that the opponent has a reason for desisting from his proposed action. If, therefore, we find a case whose conditions require that the person issuing the imperative is not the same person as the victim, actual or hypothetical, of the action, then Hare's manoeuvre cannot be brought into play. And since at least one case which he regards as of special importance may be presented as a case of this sort this particular form of argument does not appear capable of the degree of generality required of it.

A stronger defence of Hare's analysis is the following. Though it may indeed be logically impossible for his opponent to be in the particular sort of situation in which the victim of his proposed action finds himself, nevertheless it must logically be possible for him to be in a *similar* sort of situation, and this is sufficient to validate the argument against him, since his action commits him to prescribing for himself in that situation something which he does not want. Thus the South African in our example would be committed to prescribing something like this: 'Let me be treated like a chattel by anyone whose level of civilization is as much higher than mine as mine is higher than this African's'. He would, of course, protest that there is a difference between the cases, but then he is logically obliged to produce a reason for regarding this difference as a morally relevant one. An example of a morally relevant difference would be that, being less intelligent and sensitive, Africans are not distressed by being treated like chattels to anything like the extent that a white man would be; a morally irrelevant difference would be that the skin of the African is black instead of white.

Thus far the argument appears sound, but a difficulty arises over the notion of a morally relevant difference; for Hare's account of this notion (pp. 221–3) seems to reintroduce as a criterion for deciding whether or not a difference is morally relevant the question of whether or not one is prepared to legislate for oneself in a situation where the difference no longer applies. Thus, to adapt Hare's actual examples, the reason why there is a morally relevant difference between denying political freedom to Negroes and denying it to animals is that we are prepared to accept the prescription 'Let me, if an animal, be denied political liberty', but not the prescription 'Let me, if a Negro, be denied political liberty'. But, obviously, this merely reintroduces the original difficulty; for if it is logically impossible for a particular person X to be in a situation s, and situation s is differentiated from situation s′, in which it is logically possible for X to appear, by the presence in s′ of a particular feature q which is absent from s, it is clearly wrong to attempt to determine the moral relevance of q by ascertaining whether X is prepared to legislate for himself in a situation like s′ in all respects but for the absence of q. For that situation is just situation s, in which X logically cannot appear. So Hare's criterion of a morally relevant difference must be rejected.

One might well think that the remarks in these final pages of the book about the moral implications of the treatment of animals provide a sufficient indication that the theory of moral reasoning which they exemplify is in some respects faulty. On page 223 we have two examples of moral reasoning, first of all the following: 'If I were turned into an animal, I should stop having any desire for political liberty, and therefore the lack of it would be no hardship for me', and then a few lines later; 'If we were bears we should suffer horribly if treated thus; therefore we cannot accept any maxim which permits bears to be treated thus; therefore we cannot say that it is right to treat bears thus.' Now just who are 'I' and 'we' here? The supposed speaker in the first example appears to conceive that if he were turned into an animal he would be the same person as he is at the time of uttering the quoted sentence. But that supposition is clearly useless; if 'he' were an animal he would not be a person at all, far less the same person as he is at the time of uttering the sentence in question. Of course we are familiar from fairy stories with the idea of someone's being turned into an animal and yet remaining in some sense the same person as before, but if one takes the idea seriously enough to describe the situation in any detail it becomes clear that the conception is an absurd one. To see this, let us ask the speaker whether being turned into an animal involves ceasing to be a man; if not, will he really be an animal? But if so, what carries over from the man to the animal to justify the assertion that the animal is the same person as the man? Clearly, this sort of discussion is pointless; for if this were the real test of how we ought to treat animals it does not seem obvious why it should not apply to other objects as well. No one, however, is likely to take seriously the suggestion that in order to decide how we ought to treat plants it is necessary to imagine that one is a plant and then to legislate for oneself in that situation. A theory which involves that sort of fantasy has gone wrong somewhere.

These difficulties arise out of Hare's insistence that one has a reason for acting in a certain way if and only if one is prepared to prescribe that some feature or features of the situation produced by one's acting should be brought about, and a reason for not acting if and only if one is prepared to prescribe that such a feature or features should not be brought about. But in addition to the difficulties already discussed this account of having a reason poses the further problem that the sense of this notion of prescribing is itself often unclear. Thus on Hare's analysis feature f is a reason for X to do action A if and only if X assents to the prescription 'In situation s (produced by action A) let f be brought about'. Now the idea of a general prescription given in an actual situation but restricted in its application to a hypothetical situation is exemplified by the sort of notice reading 'In the event of fire, break glass and press alarm'. Here the sense of the prescription is clear, in that some actual persons, viz. those who read the notice, are being instructed or ordered by a competent authority to take certain action in a certain eventuality. But what is the analogous sense of 'In situation s let f be brought about'? The prescription here does not seem to be any ordinary sort of prescription, an order, request, entreaty, piece of advice, etc., firstly because it is not addressed to anyone in a determinate context and secondly because the object of the prescription, that f should be brought about, may not be the sort of thing which could be the intentional object of any of the above-mentioned speech-acts; one's reason for going to the play might be that one would enjoy it, but one could not prescribe in any ordinary sense of that term that one should enjoy it. This suggests, then, that this sort of prescription is *sui generis*, or more precisely that it is the sort of prescription which is assented to by anyone who accepts that the presence of f is a reason for doing A. But if this is the *full* description of this sort of prescription, just what has been explained by the account of having a reason as assenting to this sort of prescription? For the concept which was to be explained (viz. having a reason) has had to be reintroduced in order to explain the concept (viz. assenting to a prescription) which was originally introduced in order to explain *it*. So nothing has been explained at all.

Similar difficulties emerge from the account which Hare gives of this kind of prescription on pages 54–5. There he says that the prescription is the answer given by the agent to the practical question 'What shall I do?', which is not normally put into words, but given by his acting. In order, however, that it should figure in metalinguistic discussion it must be given a verbal form, for which Hare considers the form 'Let me do A' most appropriate. It is clear, then, from this and from his general account of moral reasoning that Hare regards this prescription as a verbal symbol for the action itself, so that to say that in an actual situation someone is committed to this prescription is to say that he is committed to the appropriate action. But what is it for someone to commit himself in this way to an action in a hypothetical situation? In an actual situation he commits himself to the prescription if and only if he sets about doing the prescribed action; what is the analogue to setting about the action when the situation to which the prescription applies is not actual but hypothetical? Being prepared to set about the action, perhaps; but what is the sense of 'being prepared' here? It certainly cannot mean *preparing oneself* to act, as it makes sense to do that only when the situation in question is to some degree likely to come about, whereas Hare's argument is intended to apply to hypothetical situations whose realization is admitted to be in practice impossible (pp. 93–4). Nor for the same reason, can it mean promising or undertaking to act. A more plausible suggestion might be that being prepared to set about the action in the hypothetical situation is intending, should the hypothetical situation occur, to act in the prescribed way. But in what sense can one be said to intend to act in a situation which one is certain will never occur? That appears to involve intending to do something which one is certain that one will not do, which is a very difficult notion to accept. Further, what one intends is always, ultimately, some action of one's own, but the object of the prescription is in many cases not an action of the person prescribing, but some passion of his, i.e. that he should be treated in a certain way. An example would be that discussed earlier, that of the white man who committed himself by his treatment of Africans to the prescription 'Let me be treated like a chattel by anyone whose level of civilization is as much higher than mine as mine is higher than an African's'. One can intend to allow oneself to be treated in a certain way, or even intend to make sure that one is treated in a certain way, but both these intentions go far beyond what the agent commits himself to in this case. In his relations with Africans he merely committed himself to some intention *qua* agent, and said nothing about what intentions his victims either would or ought to have. And since he might well intend to ill-treat them while recognizing that they intended not to be ill-treated (i.e. intended not to allow themselves to be ill-treated), he might with perfect consistency intend, in his new role of victim, not to allow *himself* to be ill-treated. So the analysis of committing oneself to a prescription applying to a hypothetical situation as intending to perform a certain action should that situation arise must also be rejected. Since, therefore, none of the proposed accounts of committing oneself to such a prescription has proved adequate we find ourselves once again drawn to the conclusion that committing oneself to such a prescription just is saying that in such a situation it would be right if a certain action were done, or that a certain action ought to be done. And if that is so, it appears to be otiose to suggest that someone has a reason for not saying that something ought to be done in a hypothetical situation *because* saying that commits him to prescribing for that situation the doing of an action which he is prepared to prescribe should not be done. For that analysis seems to come just to this, that one has a reason for not saying that a certain action ought to be done if one already holds that it ought not to be done. This is indeed true, but uninformative; prescriptivism seems, therefore, to have little to offer in explanation of what it is to have reasons for action.

This last difficulty clearly parallels the difficulty about *akrasia*. The problem there was that on Hare's account someone who assents to the judgement that he ought to do something must do it when the occasion arises, or else be judged either insincere or incapable of acting as he says he ought; this was seen to require an unsupported theory of psychological compulsion. In hypothetical situations, on the other hand, the sense of the judgement that one ought to do something is explained by a commitment, not indeed to action, but to a prescription whose sense is obscure. Add to these difficulties the impossibility of accounting for a morally relevant difference between cases in terms of prescriptions to which one is prepared to commit oneself and those to which one is not, and it appears that prescriptivism as outlined by Hare is left with a number of serious problems on its hands. It remains to enquire whether some modification of the theory might not be devised which would avoid these problems while preserving the strength of the prescriptivist position.

The strength of prescriptivism I take to be its emphasis on the close logical connection between evaluative judgements and action, which draws attention to such facts as the oddness of claiming that assent to the judgement that something is a good x is in no way relevant to choosing a thing of that sort when xs are to be chosen, and the need for explanation if someone, while agreeing that he ought to do something, yet fails to do it when the time for doing it comes. These advantages would be retained by a theory which explained having a reason for an action as wanting something which would be produced, directly or indirectly, by that action. On this analysis a good x would be an x which one would have a reason to choose in the appropriate circumstances, and saying that one ought to do something would be to say that one had an overriding reason for doing it. The connection with action is thus retained, since it is at least generally true that people choose the things which they think they have reasons for choosing and to do the things which they think they have the best reasons for doing. On the other hand, this analysis appears to bypass the difficulties which beset prescriptivism as presented by Hare. Thus a morally relevant difference between cases can be defined as a reason for action which applies to one case but not to the other, which is further analysed as a want which occurs in one case but not in the other. Clearly this want need not be a want of the agent who is yet said to have a further reason for acting in this case as contrasted with the other; this leaves us with the problem of saying why someone else's want ought to be a reason for anyone to act. Leaving spontaneous benevolence out of account, it is sufficient to say that to regard other people's interests as a reason for oneself to act is a requirement of acting justly, and that everyone has a reason for acting justly, since the state of affairs in which people in general act justly is the best way of ensuring the satisfaction of the wants which we are justified in assuming everyone to have. It would be plausible to suggest that someone had no reason to act justly only if his wants were such as to require for their satisfaction no co-operation from other agents; but such a life, even if logically possible, would be so restricted in its aims that it is difficult to imagine its being seriously and consistently pursued. As a second improvement on Hare's theory, we now have a better sense of assenting to a prescription in a hypothetical situation, viz. recognizing that in such a situation one would have overriding reasons for acting in a certain way. Finally, this analysis gives a better account of *akrasia*. On this account, the weak-willed man is one who, while recognizing that a certain action is one which he has the best reasons for doing, in that doing that action will agree with a general policy which he wants to follow more than he wants to follow any other policy, nevertheless fails to do the action because, when the time for action comes, he wants some temporary good, e.g. pleasure, more than he wants the long-term good of carrying out his policy. But since having a reason for doing x was analysed as wanting something which doing x would help to bring about, and having a reason for doing x rather than y as wanting the results of x more than those of y, does this account of *akrasia* not amount simply to saying that the weak-willed man decides on each occasion that he has stronger reasons for going against his policy than for following it? We have, however, a criterion for his wanting to follow his policy more than he wants to go against it, in that he does not abandon the first policy and go over to one of pursuing, e.g. pleasure as his main aim. What makes the case so puzzling is that this criterion of his wanting to follow the policy more than he wants to go against it conflicts with the criterion which is provided by his actually going against it. Then why not say that what he really wants is to live a life which combines a general adherence to his policy with frequent lapses from it? But in that case his claim to want to adhere to his policy all the time is insincere. No doubt many cases are properly to be described in this way, but we seem able to distinguish from these cases where the man 'really wants' to adhere to his policy in the strict sense, cases which are distinguished by such features as the seriousness of his efforts to adhere to the strict policy and by the degree of remorse at failing to do so. Clearly, what these features are will depend on the type of case; the man who sometimes walks across the grass while agreeing that one ought not to is not likely to make serious efforts at self-reformation or to feel any great remorse at his lapses, since the matter is not serious enough to make such attitudes appropriate. But his case can still be distinguished from that of the man who thinks that one ought not to cross the lawn unless one is in a hurry, by such features as his not claiming that being in a hurry is a justification for crossing the lawn, his agreeing to strict measures against lawn-crossers, etc. In addition to the difficulty of making this distinction there is the problem of distinguishing the man who more or less idly wishes that he were the sort of man who wanted some long-term good from the man who does want it but does not always act with a view to obtaining it, of distinguishing the man who wishes that he were the sort of chap who really would work for a First from the man who does want a First, and knows that if he wants it he should be working, and is not working. Clearly, in deciding whether to describe someone in the latter way rather than the former, the amount of work which he actually does will be of crucial importance; so Hare is right in making action in accordance with a principle the central criterion of assent to that principle. But that does not justify drawing the criteria so narrowly that anyone who really assents to a principle must be judged incapable of acting on it whenever he fails to do so. The difficulties involved in that account have already been clearly seen.

Is there any reason why Hare should be unwilling to accept this revised version of his argument? His main reason for insisting on prescriptivism is that without this theory it is impossible adequately to account for the function of moral language in guiding conduct (e.g. p. 23; *Language of Morals*, p. 172). But the best possible way of guiding someone's conduct is to show him that he has good reasons for acting in a certain way. Further, this account seems quite close to Hare's own thought, as given in *Freedom and Reason*. Thus at page 71 he points out that 'the kind of thought that we have when we want something belongs with the kinds of thought that are expressed in prescriptive language', and at page 169 he says that 'neither aesthetic preferences nor moral ideals could have the bearing that they do upon our actions unless they were prescriptive. . . . The same point, substantially, can be put in terms of *wanting* or *desiring*, if we use those words in sufficiently wide senses.' It is my contention that putting the point in the latter way avoids some difficulties which are raised by putting it in the way favoured by Hare. Again, in the account of moral argument in chapter 6, the questions of what prescriptions one can and what one can not make for oneself with respect to hypothetical situations are decided by consideration of what one wants, e.g. it is because one wants not to be tortured that one cannot prescribe that one should be tortured. So why not just say that the fact that one wants not to be tortured provides a reason for not saying that it is right to torture people? To the objection that prescriptions must be universalizable, whereas wants are not, it may be replied that universalizability is brought in by the use of the concept of a reason, since it is part of that concept that what counts as a reason for one person with respect to one situation must be a reason for all similar persons with respect to all similar situations. Thus if my wanting not to be tortured is to be a reason for other people not to torture me (a state of affairs which I want), that commits me to admitting that their not wanting to be tortured is a reason for my not torturing them, and since in fact most people don't want to be tortured it follows that I have a reason for not torturing them.

If this reformulation of prescriptivism is accepted, it then becomes interesting to enquire how wide an area of disagreement remains between Hare and his 'naturalist' opponents. For both would agree that given an object satisfies some want of an agent, then that agent has a reason for choosing that object, and also a reason for calling it a good x (for his purposes). Both would further agree that the description of an x as a good x *simpliciter* is determined not by any want of any particular person which it happens to satisfy, but by the want or set of wants which constitute the standard interest in objects of that sort, the satisfaction of which, in the case of a functional object, is identical with the function of the object. This point has in the past been emphasized more by naturalists than by prescriptivists; on the other hand the latter can fairly claim that naturalists have consistently ignored the extent to which such standard interests can be altered under the pressure of interests which were at first (rightly) regarded as deviant, and in particular that they have tended to overlook the analogous process by which standards of moral and other sorts of human excellence come to be changed. But given that both these features are admitted in the concept of a good x, and particularly in such concepts as those of a good action, a good human life and a good man, where does the dispute lie? It appears to lie in this, that Hare is committed to holding that the range of interests which could determine the application of such characteristics as 'good man' and 'good life' is indefinitely extensible, whereas naturalists hold that it is finite and in principle specifiable. Further, they appear to be committed to a distinction between 'good life' and 'good knife' on the following lines, that whereas someone might in a suitably described case properly claim to have no interest in good knives as such, but only in knives which were good for his purposes (which might well be bad knives *qua* knives), no one could sensibly claim to have no interest in the good life as such, since every individual must have at least a certain number of the wants which make up the standard interest in the good life. We can thus, and only thus, be certain that everyone has a reason for doing the actions which constitute the good life as determined by that standard interest. That Hare, at least, agrees that the residual dispute lies in this area appears from the following remark from page 110: 'It is, indeed, in the logical possibility of wanting *anything* (neutrally described) that the "freedom" which is alluded to in my title essentially consists.'

If this is in fact the nature of the dispute, a certain scepticism as to its importance may be pardoned. For if we know that the great majority of people have the standard interest in human life, then we have a sufficient basis for working out a scheme of good reasons for action of whose application to most cases we can be confident. The question of whether there might conceivably be cases to which none of these reasons applied then seems a trifle academic; for such questions resolve themselves into disputes about whether some creatures with very queer wants should properly be called men and whether their ways of life (supposing them to be coherently describable) could seriously be recommended as good sorts of human life. Yet questions of serious ethical interest arise, as Hare recognizes, not from such puzzles, but from the difficulty of deciding just which out of the immense variety of recognizably human wants are to count as constituting the standard interest in human life. An example is the difficulty of showing the error of the man who claims that the best state of humanity is reached by the attainment of absolute power by a few, irrespective of the claims of justice. Theorizing about the nature of human wants seems to be of little help here; should we not rather proceed more modestly, by a method closer to Hare's, taking generally recognized human wants as the basis of a morality which allows us to establish that justice, temperance, etc., are virtues and injustice, cruelty, etc., vices? Most of us *have* reasons for practising these virtues and avoiding these vices, and should we come across someone who appears to have no such reason, the fact that we have reasons for wanting everyone to act in this way gives us a reason for making sure that he does too.

It remains only to add that in his final chapter Hare gives a penetrating analysis of some kinds of bad argument and false assumption which are commonly produced in the attempt to justify immoral conduct. In exposing this sort of sophistry, and in expounding his general argument with such extreme fairness and precision, Hare performs a notable service to that ideal of clear thinking about morals to which his book is dedicated.

5

Two concepts of morality

NEIL COOPER[1]

It is a surprising fact that moral philosophers have rarely examined the distinction between what I shall call 'positive' or 'social' morality on the one hand and 'autonomous' or 'individual' morality on the other. Accordingly, conceptual and moral issues of the greatest importance have been neglected. The distinction is, I take it, recognized by Hegel, when he contrasts *Sittlichkeit* with *Moralität*. However, the rival sides who give a conceptual or a moral preference to one concept over the other rarely come to grips with one another, and the deep conflicts between them are concealed instead of being brought out into the open. Only in Burke's diatribe against Rousseau, Bradley's critique of Sidgwick (*Collected Essays I*, 122), Hobhouse's crusade against Bosanquet *Metaphysical Theory of the State*), Prichard's attack on Green, (*Moral Obligation*, p. 75), Hart's criticism of Hare ('Legal and Moral Obligation' in *Essays in Moral Philosophy*, edited by Melden), and above all in Oakeshott's onslaught on Rationalism (*Rationalism in Politics, passim*) do we get a glimpse of one of the main issues of moral philosophy and of morality. For just as we have two concepts, so we have two moral conceptual schemes, each of which gives a central place to one concept at the expense of the other. Those who suppose that morality is or ought to be wholly or mainly a social concept may recommend submission

[1] An earlier version of this paper was read to the Scots Philosophical Club on 16 May 1964. I am indebted for criticisms to members of the Club and also to Professor R. M. Hare and Mr L. C. Holborrow.

Reprinted with modifications from Philosophy, *1966, pp. 19–33, by permission of the author and* Philosophy.

to a tradition. Those, on the other hand, who suppose morality to be primarily an individual or independent concept will recommend independent decisions. I want in this paper, firstly, to explain the differences between the two concepts, secondly, to show that neither of them is conceptually illegitimate or degenerate, and lastly, to determine what place, if any, each ought to have in a rational morality.

Firstly, I shall consider the morality of a social group. Morality, like law and religion, is plainly a social phenomenon; unless there were such a thing as human society, there would be no such thing as morality. Just as there are objective criteria for determining what the law of a particular country is, so there are objective criteria for determining what the morality of a social group is. In the morality of a social group we encounter customs and rules of a certain kind, those customs deviation from which is met at the very least with a distinctive kind of social pressure or sanction, 'the reproaches of one's neighbours'. A statement describing the morality of a group can be as concrete and objective as a statement of what the law of a given community is. To state that it is an offence in our community to commit adultery is to state an objective fact about our social life. It is because of this resemblance between morality and law that people have often found it tempting to talk about morality as unwritten law or laws (cf. Thucyd. II, 37: *ὅσοι ἄγραφοι ὄντες αἰσχύνην ὁμολογουμένην φέρουσιν*). Where morality is found to differ from law is not in being less concrete or objective, but in being less perspicuous and less monolithic. Morality is less perspicuous than law, for there are no statute books which will enable us to determine what is morally required in a given community, nor are there generally recognized moral courts to interpret the spirit or the letter of moral precepts. Further, it is less monolithic than law, since in a large social group there may be several moralities. There will of course, be some common ground between the different moralities, a common ground which makes a common legal system and form of life possible. But it may be unclear to what extent people in fact accept the morality to which they pay lip-service, just as there is doubt in our community as to the extent to which people accept the altruistic morality consecrated by religion and tradition.

There are two points of view from which we may make statements about the law or morality of a group (see Hart, *The Concept of Law*, pp. 86–7). When, *qua* neutral observers or social anthropologists, we state what the law or morality of a community is, we are talking from what Hart calls 'the external point of view'; we are not ourselves prescribing or evaluating, we are stating facts. It is a distinctive mark of an external statement that one may make it more explicit by adding some expression like 'Among the . . .' or 'In . . .' where the gaps are used to specify a group or country (cf. Maine, *Ancient Law*, p. 134, Everyman edition: 'Among the Hindus, the instant a son is born he acquires a vested right in his father's property'). When the social anthropologist says that Navaho men are under an obligation not to look at their mothers-in-law, he is not recommending Navaho not to look at their mothers-in-law; he is saying something about the observable rule-following behaviour of the Navaho. He is saying that, if a Navaho man does look at his mother-in-law, fellow-members of his society will regard this as a reason for imposing a sanction, verbal or otherwise. It is *possible* to use the word 'ought' externally, as when Firth (*Human Types*, Mentor edition, p. 106) says quite generally of any morality that 'there may be a wide gap between the rule which ought to be followed and what is actually done'. But to use it in this way is to deprive it of its distinctively prescriptive character. The word 'ought' is typically used when we are enunciating or applying rules. If, for example, believing that Navaho are under an obligation not to look at their mothers-in-law, I am prepared to say to a Navaho, 'You ought not to look at your mother-in-law', then I am endorsing and applying a social, perhaps a moral rule of the Navaho. In this case both when I say 'Navaho are under an obligation not to look at their mothers-in-law' and when I say, 'They ought not to look at their mothers-in-law', I am speaking, if I am sincere, from the *internal* point of view, I am speaking from the point of view of somebody who is both *in* and *of* the society.

The same two points of view affect our use of words connected with social roles and institutions. All these words are internally related to the concept of obligation. Firstly, role-words may be used either externally or internally, e.g. 'father', 'lecturer', 'witch-doctor', 'pontifex', 'plumber'. When I describe what the role of a Samoan woman is, what is expected of her in her community, I am speaking externally. On the other hand, if I tell a bus-conductor that it is his job to see that the bus does not move off while the passengers are boarding it, I am plainly prescribing, I am speaking internally. Secondly, there are institution-words, words which refer to institutions and social practices, many of which involve putting oneself or others under obligations, for example, lending, borrowing, promising. It is because concepts such as these are internally related to the concepts of obligation and right that some philosophers (for example, Locke and Leibniz) have supposed that the principles of morality were necessarily true propositions, for example, 'You ought (are under an obligation) to pay your debts', 'You ought (are under an obligation) to keep your promises'. In these statements we may either regard the institution-words as used internally or as used externally. If the word 'promise' is being used internally, then it is vacuously true that you ought to keep your promises, it is a useless tautology and so not a moral judgement, since a genuine moral judgement must be synthetic. If the words are used externally, then the statements are no longer vacuous but substantial, but in this case they are no longer incontrovertible.

In his article ('How to Derive "Ought" from "Is"', *Philosophical Review*, January 1964, reprinted in *Theories of Ethics*, edited by Philippa Foot) J. R. Searle sets out an elegant and suitably qualified derivation of an apparent 'ought' from an apparent 'is'. The derivation is, however, invalid because he unwittingly vacillates between external and internal uses of the key words of his argument. The 'derivation' is as follows:

(1) Jones uttered the words 'I hereby promise to pay you, Smith, five dollars'.
(2) Jones promised to pay Smith five dollars.
(3) Jones placed himself under an obligation to pay Smith five dollars.
(4) Jones is under an obligation to pay Smith five dollars.
(5) Jones ought to pay Smith five dollars.

Now the key words in the argument are the words 'obligation' and 'promised', each of which may be used in two ways, either internally, as providing a reason for action, or externally. When used externally 'Jones is under an obligation to do X' means 'Jones is in a position in which he will be breaking a rule (moral or legal or . . .) if he does not do X'. The word 'promised' in (2) is likewise used externally if the word is defined in terms of 'putting oneself under an obligation' in the external sense of the last word. The derivation will then proceed smoothly from (1) to (2) from (2) to

(3a) Jones put himself in a position in which, if he does not pay Smith five dollars, he will be breaking a rule

and from (3a) to

(4a) Jones is in a position in which, if he does not pay Smith five dollars, he will be breaking a rule.

But here the derivation breaks down. From (4a) we can only obtain (5), viz., Jones ought to pay Smith five dollars, if we assume the Principle of Conformity:

'Jones ought not to break a rule'

and this principle is plainly not a tautology but an evaluatively potent premise.

Let us suppose, on the other hand, that the words 'obligation' and 'promised' are used internally. Here again we may glide smoothly from (2) to

(3b) Jones put himself in a position in which he will be breaking a rule he ought to keep if he does not pay Smith five dollars,

from (3b) to

(4b) Jones is in a position in which he will be breaking a rule he ought to keep if he does not pay Smith five dollars

and from (4b) to

(5) Jones ought to pay Smith five dollars.

However, in this case the breakdown occurs at the very beginning, for if 'promised' is used internally in (2), we cannot without a supplementary premise derive (2) from (1), viz., Jones uttered the words 'I hereby promise to pay you, Smith, five dollars'. This supplementary premise would be to the effect that saying the words 'I promise' constituted a good *prima facie* reason for carrying out the action contemplated. For if the word 'promised' is used internally in (2), the author or speaker is *eo ipso* endorsing, even if only tacitly, the institution of promise-keeping. This he is not committed to doing by the mere assertion of (1) where he is not *using* the word 'promise', but only quoting it. Plainly (1) could be sincerely asserted by somebody who wanted to eschew value-commitments altogether or even saw no virtue in the practice of promise-keeping at all.

Searle's argument, then, is invalid if he confines himself to one of the two kinds of uses of institutional words. The argument only derives some plausibility if Searle commits an equivocation and vacillates between external and internal uses of the key words. Searle's argument and any similar argument which makes use of role-words or institution-words to derive an 'ought' from an 'is' depends for its persuasiveness on evaluatively potent premises which are embedded in institutions and roles. For such arguments to work we have first to accept that uttering a certain form of words (e.g. saying 'I promise' or 'I challenge you'), going through a certain procedure (e.g. handing some money to a person behind a counter), standing in a certain relationship (e.g. being the son of) constitute good *prima facie* reasons for certain kinds of behaviour.

Now the distinction between external and internal uses of language is one which is made within the concept of social morality. To use a moral word internally is not *eo ipso* to make a moral judgement off one's own bat, to make an autonomous or independent moral judgement; it is to endorse or apply a moral rule of the social group. The person who speaks from the internal point of view, i.e. makes an internal use of a moral word in all sincerity, regards the rule which he applies as a reason for action. Such a person does not necessarily make an autonomous moral judgement about the rule. He may endorse and apply the rule merely out of habit, indolence, fear or unreflective respect for tradition. The internal point of view may be taken by a participant in a society's way of life even if he is critical of it and only half-hearted in his acceptance of it. On the other hand, it is possible for somebody who takes the internal point of view to make an autonomous moral judgement about the rule. He may endorse or apply the rule, not out of habit or fear, but because he thinks it a good rule or a rule which ought to be followed. Such a person, in so far as he is prepared to judge the morality of his society, whether it be favourably or adversely, may be said to have a morality of his own. It may of course even be the case that he agrees in all or almost all essentials with the morality of the society in which he lives. But since his adherence is not blind but reflective, he may be said to have a morality of his own. An individual no less than a social group may have a morality, what the individual approves or disapproves of, or thinks that he ought or ought not to do, though we talk not of an individual's moral rules, but of his moral principles and moral beliefs.

We have, then, two concepts. The first concept I call the 'positive' or 'social' concept of morality, sometimes called 'the law of opinion'. I use the expression 'positive morality' in the same sense as Austin (*The Province of Jurisprudence Determined*, Lecture 1). The second concept I call the 'autonomous', 'independent' or 'individual' concept of morality. We employ the autonomous concept whenever we stand back from the positive morality and try to make up our minds whether to accept or reject some part of it. Thus if I say 'We have moral duties and so we ought to have moral rights', the words 'rights' and 'duties' refer to the positive morality, while the word 'ought' is independent of the positive morality and thus autonomous. (I suggest that second-order moral deontic modalities are possible where the lower-order modality is positive and the higher-order one is autonomous, but not *vice versa*. Thus in the sentence 'It ought to be the case that we ought to love our enemies', the second 'ought' must be positive and cannot be autonomous.) An autonomous moral judgement is, then, a moral judgement which purports to be made from an autonomous point of view, it is a judgement made off one's own bat, which may or may not coincide with the positive morality or moralities of the social group to which the speaker belongs.

Now autonomous moral judgements admit of two different uses, a supervenient use and a non-supervenient use. An autonomous moral judgement is used 'superveniently' or 'critically' if the speaker is able to give reasons for making it, if the moral word in the judgement is applied to acts or people in virtue of their possessing certain characteristics. An autonomous moral judgement is used 'non-superveniently' if the speaker is not able to give reasons for it, if he applies the moral word in the judgement to something 'immediately', without analysis of its characteristics.[1] Someone who in all sincerity uses an autonomous moral judgement 'superveniently' or 'critically' may be said to be making an autonomous moral judgement 'from the critical point of view'. On the other hand, someone who in all sincerity uses an autonomous moral judgement 'non-superveniently' or 'immediately' may be said to be making an autonomous moral judgement 'from the immediate point of view'. Which of these two points of view someone is speaking from depends on whether or not he is ready to give reasons for his judgement.

Two concepts and four points of view have been distinguished. A comparison of the four can be instructive. The external judgement is purely descriptive, the internal judgement both descriptive and prescriptive, the critical autonomous judgement both descriptive and prescriptive, the immediate autonomous judgement purely prescriptive. The two most easily confused are the middle ones. Bentham, for example, in his *Fragment on Government* seems to vacillate between regarding moral judgements as internal, representing the voice of the people, and regarding them as autonomous, as when he stresses the importance of 'private judgement'. In either case he wants to give moral judgements objectivity. In the former case, they acquire objectivity by being tied to positive morality, in the latter case they acquire it through the existence of a common criterion for applying moral terms, namely, the Greatest Happiness Principle.

It has sometimes been maintained that autonomous moral judgements have no 'non-supervenient' or 'immediate' use. It

[1] Of course it may be possible for a spectator to discover by an *ex post facto* analysis *the* reasons for a man's making a given immediate moral judgement, but this is not the same thing as discovering *his* reasons.

has been suggested that it is part 'of the logic' of moral terms that they have a 'descriptive meaning', even though this descriptive meaning has been supposed to be secondary. Whenever anybody makes a moral judgement, he must, on this view, be applying criteria and must be 'prepared to apply them universally' (Nowell-Smith, *Ethics*, p. 167). It would seem to follow from this that anybody who used moral judgements solely to express his emotional reactions to particular situations would be misusing them and failing to make genuine moral judgements. As Marcus Singer puts it (*Generalization in Ethics*, p. 59), 'A statement having the verbal form of a moral judgement for which one is unable to give reasons does not express a genuine moral judgement at all.' This account of the correct use of moral terms I shall call the 'Criterionist' account. It is, I shall argue, mistaken.

In the first place, it may be conceded that the Criterionist account is true of the way Criterionists very often use moral words, but this hardly justifies the Criterionist claim to be giving *the* correct analysis of moral concepts. There is no denying that many people do apply moral words and hence make moral judgements without having any reasons or criteria of which they are conscious. Since any account of the actual use of moral words must cater for all uses, we have to recognize two different 'sub-logics' of moral words, both parts of the 'overall logic'. Criterionists and their anti-Criterionist opponents use moral terms differently because they have different conceptions of morality and adopt different conceptual frameworks. (The same sort of schism in usage occurs in the case of the word 'probable'. It is used without criteria on some occasions and with criteria on others, and this gives rise to a pseudo-conceptual conflict between the subjective or personalist theory and the objective or frequency theory.[1] The conflict in both cases is over the desirability or undesirability of certain uses of words, not over the 'nature of' probability or morality.)

Secondly, even the Criterionist must on some occasions employ moral words immediately and without reasons if he wishes to state explicitly what his moral beliefs are. For if he wants to state

[1] See my paper 'The Concept of Probability', *British Journal for the Philosophy of Science*, 1965.

what his *basic* criteria are ('Whatever promotes the greatest happiness is right', for example), he cannot in stating this be applying any further criteria. This follows from the meaning of the word 'basic'. In stating what his criteria are, he must be *using* the moral words as opposed to mentioning them, and if he is using them without reasons or criteria, then he must be using them in that 'immediate' sense which he has ruled out as contrary to the correct use of moral language. It follows from this that anybody who maintains the existence of critical autonomous moral judgements must concede the existence of immediate autonomous judgements also.

We are now in a position to ask some questions of a different kind. So far I have tried to describe the actual uses of moral words. However, it seems to me that the moral philosopher can and should do more than this. Some moral philosophers, while conceding the logical possibility of immediate autonomous moral judgements, have attempted to deny their propriety and have pooh-poohed them as mere expressions of personal attitudes, thus themselves prescribing rather than describing the uses of moral language. Again, others have wanted to down-grade one of the two concepts to the advantage of the other and so to present a conceptual scheme dominated by one of these concepts. I suggest that it is the task of the moral philosopher to decide whether such pooh-poohing and down-grading is justified and rational, and since criteria of rationality are not in all cases given, the moral philosopher may have to propose and defend criteria of rationality.

There are two questions I want to ask about the two concepts. The first one concerns the concept of autonomous morality. We have seen that both critical and immediate uses of autonomous moral judgements are conceptually possible and legitimate, but we must further ask which use of autonomous moral judgements is the more rational one. Secondly, if, as I shall argue, it is more rational to make moral judgements superveniently than to make them immediately, we should go on to ask whether it is rational in coming to decisions to prefer the social concept of morality to the autonomous concept, where the autonomous concept is represented by supervenient or critical autonomous moral judgements.

In order to answer either of these questions I must specify what I am going to mean by 'rational' and 'irrational'. All I need here, I think, is a minimum criterion of rationality and irrationality. To choose means which will promote one's ends or what one wants is rational; to choose means which will frustrate or fail to promote one's ends or what one wants is irrational. Further, to choose means which are more conducive to one's ends or aims is more rational than to choose those which are less so.

Now to answer the first question, whether it is more rational to make critical or immediate autonomous moral judgements, we have to consider what our aims or ends are in using moral language. If we are going to arrive at categorical rather than hypothetical judgements of rationality and irrationality here, we shall have to specify invariant ends or aims, ends which anybody who uses moral language necessarily commits himself to. If there were no such ends, we should not be able to make any categorical assertion of rationality or irrationality. For if we said that a certain non-invariant end was *the* end of moral language, this would be a prescription and not a conceptual statement, and any statement of what it was rational or irrational to do would be conditional on the acceptance of the prescription.

It is, I suggest, an invariant end of moral language to aim at agreement about what to do. True, we use moral language not only prospectively and prescriptively, but also in ways which are less imperatival, less obviously relevant to questions about what to do here and now, as when we consider what somebody ought to have done or whether somebody long since dead was a good man or not. But such uses of moral language derive their interest from their connection with moral talk which is more closely related to the question 'What are we to do?'. Whether a moral judgement about the past is actually relevant to any live question about what to do depends on how much the present resembles the past, but even a moral judgement about a temporally remote event is capable of being relevant to present action. The general purpose of using moral language is, then, to obtain agreement on what to do. For it could not be the case that our sole aim in using moral language was to express our emotional reactions, to let off steam in a peculiar way about particular topics. If that were our sole aim, we should be satisfied merely with having *expressed* our moral indignation, say, and care not one jot whether our auditors agreed with us. But in any normal interpersonal situation we do care whether what we approve or disapprove of is or is not done, and hence we must also care what our auditors' attitudes are and whether they agree with ours, since actions spring from attitudes. When we pass an autonomous moral judgement in an interpersonal situation, when, say, we call someone's proposed action 'wrong', it must be the case, if we are sincere, that we want him to avoid that action and to take up towards it an attitude similar to our own.

Now since, when we make a sincere autonomous use of the word 'wrong', we want to elicit agreement and so secure the avoidance of certain behaviour, it is more rational to use the word 'wrong' in a way which will increase rather than decrease the likelihood of agreement and consequent conformity. If in using the word 'wrong' in an interpersonal situation we do no more than express an adverse reaction to the action, we neither have nor can hope to have any hold or leverage over our auditors. For the direct confrontation of one person's emotional reactions with another's does not of itself produce agreement, indeed it may produce the reverse. Someone who makes an 'immediate' use of the word 'wrong', the non-Criterionist, invites the retort, 'Why should we take any notice of what you are saying? You are merely expressing your personal likes and dislikes.' The non-Criterionist has no answer to this retort and this is a shortcoming in his use of moral language. On the other hand, if we use the word 'wrong' as the Criterionist does, in such a way that there are criteria for its application, then there is hope of being able to exercise some leverage over our audience. For either the criteria are agreed or they are not. If they are agreed, then both speaker and audience share the same attitudes, and any disagreement is reduced to one about matter of fact. But even if there is disagreement about criteria, since criteria are being employed by both sides, each side will be aware of what universal evaluations the other side is committed to. But to understand any universal evaluation is to know what would be a counter-instance. 'Magna est vis instantiae negativae', especially in morals.

It is through the production of counter-instances that each can exercise leverage on the other. The *argumentum ad hominem* in moral discussion is the production of counter-instances acceptable to the 'hominem'. But whether what are alleged to be counter-instances are accepted or not does not matter to my argument. As long as it is possible to raise counter-examples, each side can hope to exercise leverage on the other. As long as discussion is possible, there is hope of obtaining agreement, but where there is head-on collision, there is no possibility of discussion and hence less hope of obtaining agreement. Since, then, critical uses of autonomous moral language offer greater hope of obtaining agreement than immediate uses and since agreement is an invariant aim of moral language, it is rational to prefer the critical uses to the immediate uses of moral language.

Since it is more rational to employ critical than immediate uses of autonomous moral language, we can in dealing with our second question ignore immediate uses of autonomous moral language and suppose that the case for the greater importance of the individual or autonomous concept is to be presented by the Criterionist. The question, then, which we have to consider is 'To which concept, the positive or the autonomous, is it rational to give priority in coming to moral decisions?', or, to put it differently, 'Which kind of moral judgement, positive or autonomous (critical) is it rational to regard as overriding in coming to a moral decision?' On the one side we have what we may call 'Establishment Moralists' or 'Traditionalists', who would want to say that spurious moral questions are raised because people fail to realize that the real or genuine concept of morality is the social one. As Bradley puts it:

1 'There is here no need to ask and by some scientific process find out what is moral, for morality exists all round us' (*Ethical Studies*, p. 187).
2 'To wish to be better than the world is to be already on the threshold of immorality' (ibid., p. 199).
3 'The world of our fancies and wishes, the home of absolute categorical imperatives has no place in legitimate speculation' (*Coll. Essays*, I, 122).

The Establishment Moralist or Traditionalist urges that we should take our departure from tradition and give priority to existing institutions, existing moral rules. As Burke put it (*Reflections on the Revolution in France*, p. 83, Everyman ed.), 'No discoveries are to be made in morality,' and again (p. 84), '. . . We are afraid to put men to live and trade each on his private stock of reason; because we suspect that this stock in each man is small, and that the individuals would do better to avail themselves of the general bank and capital of ages.' Our rules and institutions may not be perfect, but they should, the Traditionalist maintains, be gradually improved rather than suddenly abandoned. Imperfections will be weeded out in the course of time and institutions which are unfit to survive will of themselves decline and wither away.

On the other hand, what we may call 'Autonomous Moralists' (Bentham, the Oxford Intuitionists, Hare) or Anti-Traditionalists seem to maintain the view that it is the autonomous concept which is the prior one and that the positive concept is secondary or even degenerate. In fact, Autonomous Moralists when they undertake the analysis of moral language either examine autonomous uses of moral words exclusively or pooh-pooh non-autonomous uses of moral words as 'hollow' or 'inverted-commas' uses. That the Autonomous Moralists dominate contemporary British moral philosophy is not surprising, since their predecessors, Sidgwick, Moore, Prichard, Ross, were themselves obsessed by the autonomous uses of moral words, indeed so obsessed that they were led to give autonomous uses to such social words as 'obligation' and 'duty'. Further, the Autonomous Moralist or Anti-Traditionalist maintains that 'to accept a traditional moral standard unreflectively is to fall short in human dignity, which is only fully achieved by individuals who accept only those standards that they consciously approve of for themselves and others' (H. B. Acton, 'Tradition and Other Forms of Order', *Proceedings of the Aristotelian Society*, 1952–3, p. 25). On this view, it is absurd to object to innovation, for as Bentham puts it (*Fragment on Government*, Preface to First Edition), 'Whatever *now* is establishment, once was innovation.'

I want now to argue that neither concept is to be pooh-poohed as secondary, degenerate or bogus. There are, I think, two points to be made in defence of the social concept. Firstly, it is a mistake to suppose that the only genuine moral judgements are autonomous ones. A moral judgement which is accepted unreflectively as part of a tradition or through indolence or deference is not necessarily prescriptively dead or 'hollow'. For it can still be used internally. As long as such a moral judgement is used as a reason for action or a reason for criticism, it is prescriptive. Even a judgement expressing an 'ossified' standard may be used to prescribe.

Secondly, while it may be desirable that we should always be able to give reasons for our moral judgements, yet unless the chain of reasons came to an end, we should never learn a morality at all. Some moral judgements must then be assented to without a reason; these 'unreasoned' judgements furnish the reasons why we should accept further moral judgements. To put what I am saying in mock-Kantian terms, the existence of 'unreasoned' moral judgements is a necessary condition of the possibility of moral experience. Some 'unreasoned' moral judgements are given to us in the form of traditions, institutions and 'prejudices' in the Burkean sense. Moral thinking does not any more than other kinds of thinking take place completely *in vacuo*. The raw material of our moral thought is provided by the precepts we are taught, by the customs in which we are trained. Yet although this is true, it does not justify us in arguing that the social concept must therefore be prior to the autonomous concept, that we can only understand individual or autonomous morality 'as a development from the primary phenomenon of the morality of a social group' (Hart in 'Legal and Moral Obligation', *Essays in Moral Philosophy*, ed. Melden). On the other hand, though the social morality of our time provides us with the raw material of our moral thinking it has taken its rise from the patterns of behaviour manifested by individuals, whether these patterns of behaviour are the outcome of chance, of natural growth or of deliberation. Indeed to say that something is a rule of positive morality is to say something analysable in terms of the behaviour-patterns of individuals. If we unpack what is meant by 'positive morality' we come down to the opinions of individual people, and this indeed is implied by calling positive morality 'the law of opinion'. Prichard was making the same point in his attack on Green when he said (*Moral Obligation*, p. 75) of custom or the law of opinion: 'The phrase can only be a veiled term for the thoughts held by individuals that certain actions are duties.' (Where Prichard went wrong in this passage, I suggest, was in using the word 'duty' in too wide a sense.) Similarly when it is said that the culture selects ways of conduct, this is, to quote Kluckhohn, a 'metaphorical way of speaking. The original choice was necessarily made by one or more individuals and then followed by other individuals (or it wouldn't have become culture).' (*Mirror for Man*, Mentor edition, p. 57.) Such a 'choice' need not, however, be a reflective or deliberate choice. Further, the possibility of change in a social morality is dependent upon change in the opinions of individuals. There is no procedure for the deliberate change of the moral rules of a community; they change gradually as a result of anterior changes in the moralities of individuals. For in human societies unique and incontestable interpretations of moral precepts are not possible, and it is for this reason that, as Durkheim puts it (*Sociology and Philosophy*, p. 40), 'no individual can be completely in tune with the morality of his time'.

Now it is not the case, as Hart suggests ('Legal and Moral Obligation', op. cit., p. 100), that we can only understand the morality of the individual 'as a development from the primary phenomenon of the morality of a social group'. In the first place, there are surely at least two ways and not, as Hart implies, merely one way in which a social morality comes into being. Social morality may evolve from unreflectively accepted customs. Individuals behave in certain ways and their conduct is then consciously or unconsciously imitated by others and becomes a norm. On the other hand, a social morality may take its rise from the conscious decisions of individuals. Even in a theocratic society certain moral rules are recognized because of certain moral decisions allegedly made by a God or an inspired legislator, a Moses perhaps, who puts the fear of God into those neither able nor willing to make moral decisions for themselves. Secondly, it is surely logically possible for individuals to have autonomous moralities of their own, specifying how they are to behave to other people, anterior to the existence of any positive morality.

They might not have the concepts of Obligation and Rights, but they would be able to make intelligible to one another their conceptions of how all members of the society ought to live. The myth of a Natural Law and a State of Nature in which it holds sway is a shadow cast on political theory by this logical possibility. It seems, then, that we can explain what we mean by an 'autonomous morality' without reference to positive or social morality. I have argued that neither the positive nor the autonomous concept are degenerate and that neither is parasitic upon the other. Positive-moral judgements are just as genuine moral judgements as autonomous ones, and indeed the positive morality of a society provides the jumping-off ground for moral thinking. On the other hand, autonomous moral judgements are intelligible without positive morality and change in any positive morality is dependent on prior changes in the autonomous moralities of individuals. The two concepts are interlocked.

Since both concepts are genuine concepts of morality, we can go on to contrast two conceptual schemes, the traditionalist conceptual scheme which emphasizes the positive or social concept, custom and tradition, and recommends that we should derive our moral decisions from positive morality, and anti-traditionalism which emphasizes the autonomous concept, private judgement and reflection. I shall argue that both positions, if maintained in their extreme forms, have consequences which we ought to condemn as 'irrational' and that therefore a rational morality should contain elements of both positions. As I stated before, to choose means which will promote one's ends or what one wants is rational; to choose means which will frustrate or fail to promote one's ends or what one wants is irrational. It follows from this that if we are to show that extreme traditionalism and anti-traditionalism are both irrational, we shall have to express these positions in the language of 'wanting' and 'not wanting'. This we can do, at any rate up to a point, for the following reason. When someone says in all sincerity that something is an evil or is evil, he is speaking either from an internal or from an autonomous point of view, and so he must want to get rid of what he calls 'an evil' or 'evil'. Now the extreme anti-traditionalist will argue that if someone thinks that there is an evil in a moral rule or an institution, then he *eo ipso* wants to get rid of it, and that therefore it is *always* rational for him to try to get rid of it more quickly rather than less quickly. If we simply sit back and allow the evils to be gradually weeded out by a process of natural selection, we shall have more evil for longer. This view is superficially attractive but false. For there are situations in which the very attempt to get what one wants more quickly will produce consequences more unwelcome than the original evils. It is this truth which the moderate elements in a revolution ignore to their cost. But because it is not *always* rational to try to get rid of an evil more quickly, it does not follow that we should accept the extreme traditionalist viewpoint that it is *never* rational to try to get rid of an evil more quickly. Some of the most pressing problems of politics and of private morality are problems of determining on what occasions and to what extent the process of getting rid of evils itself produces evils and whether these evils are remediable or irremediable. To believe that we should *always* get rid of evils more quickly is irrational because it sometimes leads to the frustration of one's ends, namely in those situations where 'the cure is worse than the disease'. To believe that we should *never* take deliberate steps to get rid of evils will also sometimes lead to irrational behaviour, namely in all those situations in which it is possible to get rid of evils by speedy but relatively harmless means. Since both positions are irrational in that they commit the holder to irrational behaviour, the rational procedure is to steer a middle course. We have in every case to make up our minds which kind of situation we are in, and for this we need our 'private stock of reason'.

The argument I have sketched has one serious limitation. It only applies to the extremists on either side in so far as their positions can be expressed in the language of 'wanting' and 'not wanting'. Thus if an anti-traditionalist, a radical revolutionary perhaps, wishes to maintain that the evils of the Establishment are so great that, whatever the consequences, any means are justified which will remove the evils as quickly as possible, my argument can get no leverage on him. If and only if he concedes the possibility of his being mistaken, can he be compelled to admit that his proposed irreversible change for the better *may* turn out to be an irremediable change for the worse. Once he admits this, he must further grant, though this is the most we can extort, that it may be rational for him to prefer reversible changes to irreversible ones, even if it takes him longer to achieve his aim.

Moreover, we have at least two traditions in our culture, a traditionalist tradition and a rationalist tradition. (That Rationalism is a tradition has been stressed by Popper in *Conjectures and Refutations*, p. 121.) Hence the traditionalist is faced with a choice between the two traditions and so there is at least one choice which is not made for him. On the other hand, the anti-traditionalist cannot refuse to have anything to do with any tradition, since his rationalism is itself a tradition. Our situation is such that willy-nilly we have to start from some tradition and have to make some choice for ourselves. Indeed, our culture has been produced by the interaction of these two rival traditions and makes progress, I submit, through the tension between them. Just as our actual penal system is a product of different concepts of punishment, so our actual social and moral life is a product of the union of and the tensions between two diverse but related concepts of morality. We should not be made to choose between the two concepts, we should not be presented with a rigid 'either/or'. Whether we speak as conceptual analysts or as moral agents, when we are asked to choose between the two concepts, the right answer and the rational answer to give is the child's answer, 'We want them both'.

6

Morality and importance

NEIL COOPER

We are sometimes told that a certain issue is a '*moral* issue'. Apparently we are to understand by this that the issue in question is fundamental and of great importance and seriousness, that it is not, for example, a mere matter of taste or manners. (Hence manners have been called 'les petites morales' or 'the smaller morals'.) If, for example, somebody breaks a *moral* rule, then his action, unlike a mere breach of custom or decorum, is to be regarded seriously. On this view, part of what we mean by 'moral' is 'important', and thus if you want to know what a man's moral values and principles are, you should study what he thinks to be most important. On the other hand, we also find it maintained that morality is just one among many social phenomena, interests, activities, aspects of our life, any of which may be elevated by different people at different times to a position above that of morality. On this view, there is no essential connection between morality and importance.

In an article entitled 'Akrasia and Criteria of Assent to Practical Principles' (*Mind*, July 1956, pp. 400–7), Professor C.K. Grant challenges the former view and maintains (p. 406) that 'it simply cannot be assumed without argument that nothing can be rationally regarded as more important than morally righteous conduct'. I want in this paper to draw attention to considerations in favour of the thesis which Grant challenges, and to show that the issue between Grant and his opponents can be clarified, if not

---

A revised version of an article originally published as a discussion note in Mind, *1968, reprinted by permission of the author and* Mind.

dissolved, by making a distinction between two concepts of morality.[1]

In order to arrive at a decision on this issue I shall look at a test case. I shall examine what I call 'monolithic theories', by which I mean 'theories which assign supreme importance to some single activity or interest to the exclusion of all others'. There are many examples of monolithic theories. The militarist considers that making war is the most important activity. The capitalist thinks that the most important thing is to produce and acquire wealth. The Machiavellian thinks that the interests of the state should be paramount. Aestheticists hold that Art is more important than anything else and that the appreciation and production of works of art are the most worthwhile human activities. Lastly, we should not forget that some thinkers may put morality highest; morality, they say, is more important than anything else. All these theories are monolithic in that if generally put into action they would bring it about that 'political institutions and social arrangements and intellect and morality and religion are crushed into a mould made to fit one activity' (R. H. Tawney, *The Acquisitive Society*, Penguin Edition, p. 43). I want to suggest that these monolithic theories are themselves moralities, in the autonomous sense of the word, and that part of the function of an autonomous morality is to provide an answer to the question 'What order of importance should be assigned to our activities and interests?'

Now these monolithic theories would properly be called 'moralities' if morality was indeed essentially connected with importance. But if morality is merely one aspect of life which some of the monolithic theorists I have mentioned happen to put lower in their order of priorities, then monolithic theories should not be called 'moralities' but something else, say, 'life-guides'. Indeed those thinkers who say 'Morality is important' imply by so saying that it is worth shouting from the house-tops and therefore is presumably not a necessary truth. But if part of what we *meant* by 'moral' were 'important', it would be a necessary truth that morality was important and so no moral point would be served by emphasizing the importance of morality.

Our 'test' problem, then, is whether we can properly call a

[1] See 'Two Concepts of Morality' (reprinted here).

monolithic theory a 'morality'. Someone might at the very start object even to our problem. 'Why should we bother ourselves over footling questions about the meaning of words? How typical of British Futilitarianism! What on earth does it matter how the word "morality" is used?' On this view such undue attention to words might blunt rather than 'sharpen our awareness' of the phenomena with which we are concerned. Certainly the words 'moral' and 'morality' are words of our ordinary language, but they are also the kind of words in which philosophers are especially interested, 'universe-of-discourse' words of which 'legal', 'aesthetic', 'scientific' are other examples. The philosopher's task is *inter alia* a classificatory one, that of examining the different kinds of knowledge, belief and discourse which we employ, and he may on occasion need to refine these 'tool-words' and give them a meaning which they did not have in ordinary language. Whether or not this will be necessary is a matter the philosopher has to decide in the light of his knowledge of ordinary usage. In the present case, when we ask whether there is an essential connection between morality and importance, we find that ordinary usage pulls us in two different directions, and we should therefore suspect the presence of an ambiguity. I shall argue in what follows that our perplexities about morality and importance can be dispelled by distinguishing two different but related concepts of morality.

In the first place, there is plainly a sense in which morality is a department of life. When we study a culture, we attend to the different departments of life which make up that culture, a people's law, religion, morality and so on. In such contexts the word 'morality' is used to refer to a social phenomenon, a positive morality, a framework of rules and ideals, conformity to which is enforced by a distinctive kind of social pressure or sanction, 'the reproaches of one's neighbours'. In this sense of the word a person may hold that morality is less important than, say, religion. Further he may from time to time change his mind about the order of importance of these social phenomena. The rules of a positive morality are independent of the individual's attitudes and preferences and so the body of rules, standards and ideals which constitute a positive morality is relatively stable.

Because of this a positive morality can itself become the object of attitudes; it is possible for the individual to fluctuate in his attitude towards it and to come to think that something else is more important than morality. In the limiting case a person may come to regard morality in this sense as of no importance at all and cease to regard the rules of the positive morality of his society as furnishing even *prima facie* reasons for action. Perhaps this is the condition of Tawney's militarists (op. cit., p. 45) who 'resent any activity which is not coloured by the predominant interest because it seems a rival to it. So they destroy religion and art and morality.' Again, some of the great Immoralists have described themselves as going 'beyond good and evil', when they were rejecting the positive morality of their era and culture. Gide in his *Journals* (Pléiade Edition, Vol. I, p. 953) writes: 'L' idée du bien, confortable, rassurante et telle que la chérit la bourgeoisie, invite à la stagnation, au sommeil. Je crois que souvent, *le mal* . . . est d'une plus grande vertu éducatrice et initiatrice – que ce que vous appelez *le bien*.' Gide's own morality is revealed by his praise of 'le mal' and his criticism of other people's conceptions of 'le bien'. He is not rejecting the whole concept of 'le bien' but 'ce que vous appelez le bien', the positive morality of his society.

A man will not be contradicting himself, then, if he regards some other activity, interest or aspect of life as more important than the positive morality of his society, if he regards some other value as more important than morally right conduct, where he means by the expression 'morally right conduct', conduct required by the positive morality of the society of which he is a member.

He may regard the positive morality of his society from what Hart calls 'the internal point of view' (H. L. A. Hart, *The Concept of Law*, pp. 86–7), that is, as providing reasons for action. However, the internal point of view does not commit him to regarding the morality of his society as providing conclusive or overriding reasons. If he conforms to the positive morality of his society primarily through force of habit or respect for tradition, he may regard that morality as furnishing *prima facie* reasons only, which may be overriden, if he so decides, by other considerations.

The second sense of the word 'morality' is that in which it is used when we talk, for example, of 'Gide's own morality'. For we not only talk of the morality of a group, but also of the moral beliefs and principles of individuals. In this 'autonomous' sense of the word 'morality', a person's morality cannot be independent of his attitudes and preferences. The main question to which an autonomous morality provides an answer is, I suggest, 'What considerations ought we to hold most important in making up our minds what to do?', or, to use a useful expression of Matthew Arnold's (*Culture and Anarchy*, Chapter 5), 'What ought our master-concern(s) to be?'. In this sense of the word Grant's opponents are right; nobody can consistently think that anything is more important than *his* own morality, and so, to put it in an 'Irish' fashion, if anybody does think something more important than morality, then this too is part, and the most important part, of his morality. That this view is true of autonomous morality I shall try to show by means of several examples. Aestheticists think that one human activity, Art, is more important than any other. We may describe their view in two different ways. We could describe them as 'subordinating' positive morality to Art. But we could also say that their own autonomous morality consisted in putting Art first. As Gaunt says of the Bohemians, 'They had one law, one morality, one devotion and that was Art' (*The Aesthetic Adventure*, Penguin Books, p. 14). Again, Machiavelli may be described as 'subordinating' morality to prudence or expediency. This has indeed been the usual account. But in fact he sometimes uses the language of autonomous morality. For example, if he were talking the language of positive morality, one would expect him to say, 'While we are morally bound to keep our promises, it is more important to preserve oneself and therefore one ought not invariably to do what one is morally bound to do.' What he in fact says is that as men are bad and 'would not observe their faith with you, so you are *not*[1] bound to keep faith with them' (*The Prince*, World's Classics Translation, p. 78). It seems from his way of expressing his thought that he is not so much allowing considerations of expediency to override morality as converting them into overriding moral considerations, which

[1] My italics.

would excuse the Prince from being truthful, kind and so on. This interpretation is supported by Bacon who describes Machiavelli's maxims as 'these *dispensations*[1] from the laws of charity and integrity'. Further, those who hold a Capitalist morality may be described as maintaining that commercial success is more important than (positive) morality. In fact, however, capitalist moralists, if we may believe Tawney (op. cit. pp. 182–3), hold a morality in which they give priority to commercial success. 'The ideal,' he says, 'that conduct which is commercially successful may be morally wicked is as unfamiliar to the modern world as the idea that a type of social organization which is economically efficient may be inconsistent with principles of right.'

Further, there are religious thinkers who hold that religious considerations are more important than any others. Their views too may be described in two different ways. They may be described as by Grant (*Mind*, July 1956, p. 405) as holding 'not only that the dictates of morality can conflict with Divine Commandments but that when they do morals must give way'. Here 'morals' refers to the positive morality regarded from the internal point of view. Alternatively they may be regarded as holding that religious considerations *are* moral considerations which ought to override all others. Grant takes the example of Kierkegaard's *Fear and Trembling* (Translated by Lowrie, Princeton, 1941). Kierkegaard talks at length of 'a teleological suspension of the ethical' (pp. 79 ff), meaning, I suggest, by 'the ethical' much the same as 'positive morality', Hegel's 'Sittlichkeit'. If he was in fact maintaining that religious considerations overrode moral considerations, one would expect him to say of Abraham's intended sacrifice of Isaac that, although it was morally wrong, it was right from the religious point of view, and it was the latter point of view which had priority. What, however, he appears to be holding (p. 92) is that Abraham's faith is morally relevant and that because of it he was not guilty of sin in being ready to sacrifice Isaac. In this passage, then, Kierkegaard seems to be advocating a distinctive and monolithic morality, a 'Put God first' morality.

Grant and his opponents, then, have been arguing at cross-

[1] My italics.

purposes. While he has been dealing with the social or positive concept of morality, they have been speaking the language of autonomous morality. Neither party to the dispute has recognized that there are two different ways in which monolithic theories may be described, corresponding to the two senses of the word 'morality' which I have distinguished. Confusion has also been caused because the task of describing a monolithic theory is linguistically embarrassing. Their linguistic embarrassment is similar to that experienced by Spinozists. Their opponents maintained that because they identified God and Nature they were atheists. The Spinozists retorted with equal justification that they were not atheists but 'acosmists', it was not God they denied but the universe. Similarly, since monolithic theories appear to make morality coalesce with some other human activity, it is tempting to regard morality as having been 'subordinated', 'enveloped' or even 'destroyed' by the other activity. But, as I have tried to show by means of examples, it is possible for monolithic theories to be expounded in the language of autonomous morality. In this autonomous sense of the word 'morality', to think one activity, interest or aspect of life more important than some other, even if that other is the positive morality of one's society, is to commit oneself to a moral position.

Morality in the autonomous sense is not one competing interest or aspect of life amongst many others. We have to make decisions, which may properly be called 'moral decisions', on the order of priority in which we should place these aspects and activities. Many of us are not monolithic theorists, we do not pay homage to a single architectonic value. In coming to moral decisions we take into account a number of our relevant values and ends, and amongst these, at any rate for most of us, is the positive morality of the society in which we live. The attitude we adopt towards it determines what our own autonomous morality will be. If I am right, then, an autonomous morality provides an answer to the question, 'In what order of importance should our activities, interests, etc., be placed?'. The language of morals in this sense is 'the language of priorities'.

7

Social morality and individual ideal

P.F.STRAWSON[1]

Men make for themselves pictures of ideal forms of life. Such pictures are various and may be in sharp opposition to each other; and one and the same individual may be captivated by different and sharply conflicting pictures at different times. At one time it may seem to him that he should live – even that *a man* should live – in such-and-such a way; at another that the only truly satisfactory form of life is something totally different, incompatible with the first. In this way, his outlook may vary radically, not only at different periods of his life, but from day to day, even from one hour to the next. It is a function of so many variables: age, experiences, present environment, current reading, current physical state are some of them. As for the ways of life that may thus present themselves at different times as each uniquely satisfactory, there can be no doubt about their variety and opposition. The ideas of self-obliterating devotion to duty or to the service of others; of personal honour and magnanimity; of asceticism, contemplation, retreat; of action, dominance and power; of the cultivation of 'an exquisite sense of the luxurious'; of simple human solidarity and co-operative endeavour; of a refined complexity of social existence; of a constantly maintained and renewed affinity with natural things – any of these ideas, and a great many others too, may form the core and substance of a personal

[1] This paper has been read at philosophical societies in a number of British universities. I am grateful to my critics on these occasions for forcing me to make myself at least a little clearer.

Reprinted from Philosophy, *1961, pp. 1–17, by permission of the author and* Philosophy.

ideal. At some times such a picture may present itself as merely appealing or attractive; at others it may offer itself in a stronger light, as, perhaps, an image of the only sane or non-ignoble human reaction to the situation in which we find ourselves. 'The nobleness of life is to do thus' or, sometimes, 'The sanity of life is to do thus': such may be the devices with which these images present themselves.

Two quite different things may be urged against, or in mitigation of, this picture of a multiplicity of pictures. First, it might be said that the many, apparently conflicting pictures are really different parts or aspects, coming momentarily into misleading prominence, of a single picture; this latter being the composite ideal image of our coolest hours, in which every god is given his due and conflict is avoided by careful arrangement and proper subordination of part to part. And it may be true of some exceptional individuals that they entertain ideal images which exhibit just such a harmonious complexity. I believe this to be rarer than we sometimes pretend; but in any case to describe this situation is not to redescribe the situation I have spoken of, but to describe a different situation. The other mitigating point has more weight. It is that, however great the variety of images which dominate, at one time or another, our ethical imaginations, our individual lives do not, as a matter of fact, exhibit a comparable internal variety. Indeed they scarcely could. Something approaching consistency, some more or less unsteady balance, is usually detectable in the pattern of an individual person's decisions and actions. There are, so to speak, empirical grounds for ordering his ideal images in respect of practical efficacy, even, perhaps, for declaring one of them to be practically dominant. This point I shall grant. I think it is easy to exaggerate it; easy to exaggerate the unity of the personalities of those we say we know when we really know them only in one or two particular connections; easy to dismiss as phases or moods whatever lacks conformity with our only partly empirical pictures of each other. But I shall not dwell on this. What I shall dwell on is precisely this readiness, which a great many people have, to identify themselves imaginatively at different times with different and conflicting visions of the ends of life, even though these visions may receive the scantiest expression in their actual behaviour and would call for the most upsetting personal revolutions if they received more.

This fact about many people – a fact which partly explains, among other things, the enormous charm of reading novels, biographies, histories – this fact, I say, has important consequences. One consequence is that when some ideal image of a form of life is given striking expression in the words or actions of some person, its expression may evoke a response of the liveliest sympathy from those whose own patterns of life are as remote as possible from conformity to the image expressed. It is indeed impossible that one life should realize all the ideal pictures which may at one time or another attract or captivate the individual imagination. But the owner of one life may with perfect practical consistency wish that his conflicting images should all be realized in different lives. The steadiest adherence to one image may co-exist with the strongest desire that other and incompatible images should have their steady adherents too. To one who has such a desire, any doctrine that the pattern of the ideal life should be the same for all is intolerable; as it is to me. The way in which I have just expressed the position makes its practical consistency look more simple than it is. One cannot simply escape the conflict between different ideal images by diffusing their realization over different lives. For different lives interact and one's own is one of them; and there may be conflict in the areas of interaction. One is not forced to welcome this, though one may; it is simply something that in fact goes with the fulfilment of the wish for this kind of diversity in the pursuit of ends. Equally one is not precluded from taking one side in a conflict because one has wished that both sides should exist and has some sympathy with both.

I think there can be no doubt that what I have been talking about falls within the region of the ethical. I have been talking about evaluations such as *can* govern choices and decisions which are of the greatest importance to men. Whether it falls within the region of the moral, however, is something that may be doubted. Perhaps the region of the moral falls within it. Or perhaps there are no such simple inclusion-relations between them. The question is one I shall come back to later. I should like first to say something more about this region of the ethical. It could also be characterized as a region in which there are truths which are incompatible with each other. There exist, that is to say, many profound general statements which are capable of capturing the ethical imagination in the same way as it may be captured by those ideal images of which I spoke. They often take the form of general descriptive statements about man and the world. They can be incorporated into a metaphysical system, or dramatized in a religious or historical myth. Or they can exist – their most persuasive form for many – as isolated statements such as, in France, there is a whole literature of, the literature of the maxim. I will not give examples, but I will mention names. One cannot read Pascal or Flaubert, Nietzsche or Goethe, Shakespeare or Tolstoy, without encountering these profound truths. It is certainly possible, in a coolly analytical frame of mind, to mock at the whole notion of the profound truth; but we are guilty of mildly bad faith if we do. For in most of us the ethical imagination succumbs again and again to *these* pictures of man, and it is precisely as truths that we wish to characterize them while they hold us captive. But these truths have the same kind of relation to each other as those ideal images of which I have already spoken. For pictures of the one kind reflect and are reflected by pictures of the other. They capture our imagination in the same way. Hence it is as wholly futile to think that we could, without destroying their character, sytematize these truths into one coherent body of truth as it is to suppose that we could, without destroying their character, form a coherent composite image from these images. This may be expressed by saying that the region of the ethical is the region where there are truths but no truth; or, in other words, that the injunction to see life steadily *and* see it whole is absurd, for one cannot do both. I said I would give no examples, but I will allude to one near-contemporary one. Many will remember the recorded encounter between Russell and Lawrence, the attempt at sympathy and the failure to find it. That failure is recorded in such words as: 'I thought there might be something in what he said, but in the end I saw there was nothing' on the one hand; and 'Get back to mathematics where you can do some good; leave talk about human beings alone' on the other. The clash was a clash of two irreconcilable views of man, two irreconcilable attitudes. The spectator familiar with both may say: Russell is right; he tells the truth; he speaks for civilization. He may also say; Lawrence is right; he tells the truth; he speaks for life. The point is that he may say both things. It would be absurd to hope for a reconciliation of the two conflicting attitudes. It is not absurd to desire that both should exist, in conflict.

The region of the ethical, then, is a region of diverse, certainly incompatible and possibly practically conflicting ideal images or pictures of a human life, or of human life; and it is a region in which many such incompatible pictures may secure at least the imaginative, though doubtless not often the practical, allegiance of a single person. Moreover this statement itself may be seen not merely as a description of what is the case, but as a positive evaluation of evaluative diversity. Any diminution in this variety would impoverish the human scene. The multiplicity of conflicting pictures is itself the essential element in one of one's pictures of man.

Now what are the relations between the region of the ethical and the sphere of morality? One widely accepted account of the latter is in terms of the idea of rules or principles governing human behaviour which apply universally within a community or class. The class may be variously thought of as a definite social group or the human species as a whole or even the entire class of rational beings. It is not obvious how these contrasting conceptions, of diversity of ideal and of community of rule, are related to each other; and in fact, I think, the relationship is complicated. One way of trying to harmonize the ideas would be as follows. This way is extremely crude and inadequate, but it may serve as a starting point. It is obvious that many, if not all, of the ideal images of which I spoke demand for their realization the existence of some form of social organization. The demand is in varying degrees logical or empirical. Some ideals only make sense in a complex social context, and even in a particular kind of complex social context. For others, some complexity of social organization seems, rather, a practically necessary condition of the ideal's being realized in any very full or satisfactory way. Now it is a condition of the existence of any form of social organization, of any human community, that certain expectations of behaviour on the part of its members should be pretty regularly fulfilled: that some duties, one might say, should be performed, some obligations acknowledged, some rules observed. We might begin by locating the sphere of morality here. It is the sphere of the observance of rules, such that the existence of some such set of rules is a condition of the existence of a society. This is a minimal interpretation of morality. It represents it as what might literally be called a kind of public convenience: of the first importance as a condition of everything that matters, but only as a condition of everything that matters, not as something that matters in itself.

I am disposed to see considerable merit in this minimal conception of morality. By this I mean not that it is really, or nearly, an adequate conception – only that it is a useful analytical idea. There would be objections to claiming that it was an adequate conception. One objection might be simply expressed by saying that, after all, being moral is something that does matter in itself, that it is not simply an affair of complying with rules in a situation where the observance of some such rules is an indirect condition of approximating to ideal forms of life. There is a lot in this objection, But it is not an objection to *using* the minimal idea of morality. We might for example argue that there was an intricate interplay between ideal pictures of man on the one hand and the rule-requirements of social organization on the other; and that one's ordinary and vague conception of morality was the product of this interplay. This would be one way – I do not say the right way – of using the minimal idea of morality to try to get clearer about the ordinary idea. I shall come back later to this question too.

Meanwhile there is another objection to be considered. I think there is something in it as well, but that what there is in it is not at all straightforward. It turns on the idea of the universal applicability of moral rules. The idea is that it is a necessary requirement of a *moral* rule that it should at least be regarded as applying to all human beings whatever. Moral behaviour is what is demanded of men as such. But we can easily imagine, and even find, different societies held together by the observance of sets of rules which are very different from each other. Moreover we can find or imagine a single society held together by a set of rules which by no means make the same demands on all its members, but make very different demands on different classes or groups within the society. In so far as the rules which give cohesiveness to a society are acknowledged to have this limited and sectional character, they cannot, in the sense of this objection, be seen as moral rules. But the rules which do give cohesiveness to a society may well have this character, whether acknowledged or not. So the prospect of explaining true morality in terms of what I called the minimal conception of morality is a poor one. Now it is possible to admit the principle of this objection, and then meet it with a formal manœuvre. Thus a rule which governs the professional behaviour of Samoan witch-doctors can be said to apply to all men under the condition that they are witch-doctor members of a society with the general characteristics of Samoan society. Or again, a rule which might be held to apply to ten-year-old children, namely that they should obey their parents in domestic matters, could be represented as applying to all men without exception, under the condition that they were ten-year-old children. Obviously there is a certain futility about this manœuvre, and equally obviously there is no compulsion to execute it. We might simply drop the idea of moral rules as universally binding on men as men. Or we might say that though there was something in this idea, it was absurd to try to apply it directly and in detail to the question of what people were required to do in particular situations in particular societies. And here we might be tempted by another manœuvre, which we should note as a possible one even if we do not think that it, either, is altogether satisfactory. We might be tempted to say that the relevant universally applicable, and hence moral, rule, was that a human being should conform to the rules which apply to him in a particular situation in a particular society. Here universality is achieved by stepping up an order. A man should perform the duties of his station in his society. This allows for an indefinite variety of societies and of stations within them; and would also seem to allow us, in so far as we regarded the universal rule as a truly moral one, to see at least part of true morality as resting upon and presupposing what I called the minimal social interpretation of morality.

Enough, for the moment, of objections to this minimal idea. Let me set out some of its merits. First we must be clearer about what this minimal interpretation is. The fundamental idea is that of a socially sanctioned demand made on an individual in virtue merely of his membership of the society in question, or in virtue of a particular position which he occupies within it or a particular relation in which he stands to other members of it. I spoke of rules in this connection; and the rules I meant would simply be the generalized statements of demands of this type. The formula I employ for the fundamental idea is deliberately flexible, the notions of a society and of social sanctioning deliberately vague. This flexibility is necessary to do justice to the complexities of social organization and social relationships. For instance, we can regard ourselves as members of many different social groups or communities, some of which fall within others; or again, when I speak of the social sanctioning of a demand which is made on an individual member of a group in virtue of his position in the group, we may think of the social sanction of that demand sometimes as arising only within the limited group in question, sometimes as arising also within a wider group which includes that limited group. A position in a society may or may not also be, so to speak, a position in society. Thus a position in a family generally gives rise to certain demands upon the holder of that position which are recognized both within the family and within some wider group or groups within which the family falls. The same may be true of membership of a profession or even of a professional association. On the other hand, some of the demands of certain class or caste moralities receive little or no extraneous reinforcement from the wider social groupings to which the members of the limited class also belong. Or again what one might call the internal morality of an intimate personal relationship may be as private as the relationship itself. One of the merits I should claim for this approach to morality is precisely that it so easily makes room for many concepts which we habitually employ, but which tend to be neglected in moral philosophy.

Thus we talk of medical ethics, of the code of honour of a military caste, of bourgeois morality and of working-class morality. Such ideas fit more easily into an account of morality which sees it as essentially, or at any rate fundamentally, a function of social groupings than they do into the more apparently individualistic approaches which are generally current.

Another merit which I shall claim for the present approach is that it makes it relatively easy to understand such notions as those of conscientiousness, duty and obligation in a concrete and realistic way. These notions have been treated almost entirely abstractly in moral philosophy in the recent past, with the result that they have come to some of our contemporaries[1] to seem to be meaningless survivals of discarded ideas about the government of the universe. But as most ordinarily employed I do not think they are that at all. There is nothing in the least mysterious or metaphysical in the fact that duties and obligations go with offices, positions and relationships to others. The demands to be made on somebody in virtue of his occupation of a certain position may indeed be, and often are, quite explicitly listed in considerable detail. And when we call someone conscientious or say that he has a strong sense of his obligations or of duty, we do not ordinarily mean that he is haunted by the ghost of the idea of supernatural ordinances; we mean rather such things as this, that he can be counted on for sustained effort to do what is required of him in definite capacities, to fulfil the demand made on him as student or teacher or parent or soldier or whatever he may be. A certain professor once said: 'For me to be moral is to behave like a professor.'

Suppose we now raise that old philosophical question: What interest has the individual in morality? The question may force us to a more adequate conception of morality than the minimal interpretation offers by itself. It certainly forces us to strike, or to try to strike, some delicate balances. The only answer to the question so far suggested is this: that the individual's ethical imagination may be captured or fired by one or more ideal pictures of life which require for their realization the existence of

[1] Cf. G. E. M. Anscombe, 'Modern Moral Philosophy', *Philosophy*, 1958 (reprinted in this volume).

social groupings and social organizations such as could not exist in the absence of a system of social demands made on individual members of these groups or organizations. I have already hinted that this answer is too crude, that the interplay between ethical ideal and social obligation is more intricate than it suggests. The answer is also not crude enough. The picture of the ideal form of life and the associated ethical vision of the world tend to be the products of the refined mind and relatively comfortable circumstances. But when we ask what the interest of the individual is in morality, we mean to ask about all those individuals on whom socially sanctioned demands are made; not just about the imaginatively restless and materially cosy. We need not, perhaps, insist upon just the same answer for all; but, if we take the question seriously, we must insist on *some* answer for all. There may seem to be a broader answer which does not altogether depart from the form of the over-refined answer. For who could exist at all, or pursue any aim, except in some form of society? And there is no form of society without rules, without some system of socially sanctioned demands on its members. Here at least is a common interest in morality as minimally conceived, an interest which can be attributed to all those about whom the question can be raised. Still we may feel that it is not enough. And in this feeling is the germ of the reason why the minimal conception of morality is inadequate to the ordinary notion, at least in its contemporary form; and perhaps, in uncovering the reason for this inadequacy, we may discover too what there is in the notion of the universal applicability of moral rules.

We have arrived at the fact that everyone on whom some form of socially sanctioned demand is made has an interest in the existence of some system of socially sanctioned demands. But this fact seems inadequate to answer the question what the individual's interest in morality is. We can begin to understand this inadequacy by thinking of the different things that might be meant by the social sanctioning of a demand. 'Sanction' is related to 'permission' and 'approval'; and also to 'power' and to 'penalty'. A socially sanctioned demand is doubtless a demand made with the permission and approval of a society; and backed, in some form and degree, with its power. But the idea of a society as the totality of individuals subject to demands may here come apart from the idea of society as the source of sanction of those demands. The sanctioning society may simply be a sub-group of the total society, the dominant sub-group, the group in which power resides. Mere membership of the total society does not guarantee membership of the sanctioning part of the society. Nor does a mere interest in the existence of some system of socially sanctioned demands guarantee an interest in the particular system of socially sanctioned demands to which one is subjected. But unless at least one, and perhaps both, of these non-guaranteed conditions is satisfied, it does not seem that the fulfilment of a socially sanctioned demand comes anywhere near being what we should regard as the fulfilment of a moral obligation. That is to say, if I have no foothold at all in the sanctioning part of society, and if no interest of mine is safe-guarded by the system of demands to which I am subject, then, in fulfilling a demand made upon me, I may indeed, in one sense, be doing what I am obliged to do; but scarcely what I am *morally* obliged to do. No wonder, then, that the question 'What is the individual's interest in morality?' is not answered by mentioning the general interest in the existence of some system of socially sanctioned demands. The answer now scarcely appears to touch the question.

Suppose, then, that we consider the idea of a society such that all its members have *some* interest, not merely in there being a system of socially sanctioned demands, but in the actual system of demands which obtains in that society. It seems that we can ensure such an interest even to the powerless and enslaved by stipulating that the system includes not only demands made on them in the interest of their masters, but also demands made on their masters in their interests. We might be tempted to say that by thus securing to them an interest in the system of demands, we secure to them also some sort of position or foothold in the sanctioning part of society. Certainly, when the master recognizes moral obligations to his slave, we shall be at least one step nearer to allowing that the slave is not merely subject to the demands of his master, but may recognize a moral obligation to fulfil them. Even in this extreme case, then, we can approach the situation which everyone would agree to regard as characteristically moral, the situation in which there is reciprocal acknowledgement of rights and duties.

Still I think we must admit a distinction of two stages in this approach to the characteristically moral situation. Interest in claims on others and acknowledgement of claims on oneself are connected but not identical. It is a tautology, though not an easy one, that everyone subject to moral demands has some interest in morality. For a demand made on an individual is to be regarded as a moral demand only if it belongs to a system of demands which includes demands made on others in his interest. It would be agreeable, as I just now suggested, to be able to argue strictly that this fact carries with it the conclusion that mere self-conscious membership of a moral community implies at least in some degree extending one's sanction to its system of demands, to the extent of genuinely acknowledging as obligations at least some of the claims which others have on one, even if only provisionally and with the strongest desire that the system should be different. But to argue so would be to equivocate with the phrase 'membership of a moral community'. There would be nothing self-contradictory about the idea of one who recognized his interest in the system of moral demands and resolved merely to profit by it as much as he could, fulfilling its demands on himself only in so far as his interest calculably required it. He might get away with it successfully if he were subtle enough in his practice of the hypocrisy which this policy would necessarily involve. But it is an important fact that hypocrisy would be necessary. It is connected with the further fact, a fact of human nature which can probably be explained in a number of ways, that quite thoroughgoing egotism of this kind is rare. But for this fact there could be no such thing as a system of moral demands. We cannot argue that it is a tautology that *anyone* subject to moral demands who recognizes his interest in the system of demands must also genuinely acknowledge some obligations under the system. But we can argue that it is a tautology that the *generality* of those subject to moral demands must genuinely recognize some obligations under the system of demands. For if this were not so, there would be no such thing as a system of moral demands and hence no such thing as being subject to a moral demand.

These steps from a minimal to a more adequate conception of morality (i.e. to a conception which at least begins to square with what we nowadays vaguely understand by the word) may easily encourage abstract exaggerations and distortions in moral philosophy. For instance, the necessary truth that the members of a moral community in general acknowledge some moral claims upon them may be exaggerated into the idea of a self-conscious choice or adoption of the principle of those claims. So everyone appears, grandly but unplausibly, as a moral self-legislator. This is an exaggeration which has appealed, in different forms, to more than one philosopher. Again these steps reveal something genuinely universal in morality: the necessary acceptance of reciprocity of claim. And *one* way in which a demand made on one individual in the interest of others can be balanced by a demand made on others in his interest is through the operation of a general rule or principle having application to all alike. But it does not follow from this that *all* moral claims have, or are seen by those who acknowledge them as having, the character of applications of universal principles holding for all men. There is no reason why a system of moral demands characteristic of one community should, or even could, be found in every other. And even with a single system of reciprocal claims, the moral demand may essentially *not* relate to a situation in which any member of the system could find himself *vis-à-vis* any other. Here are two reasons why it is misleading to say that moral behaviour is what is demanded of men as men. It might, in some cases, be essentially what is demanded of Spartans by other Spartans, or of a king by his subjects. What is universally demanded of the members of a moral community is something like the abstract virtue of justice: a man should not insist on a particular claim while refusing to acknowledge any reciprocal claim. But from this formally universal feature of morality no consequences follow as to the universality of application of the particular rules in the observance of which, in particular situations and societies, justice consists.

One must beware, however, of meeting exaggeration with counter-exaggeration. It is important to recognize the diversity of possible systems of moral demands, and the diversity of demands which may be made within any system. But it is also important to recognize that certain human interests are so fundamental and so general that they must be universally acknowledged in some form and to some degree in any conceivable moral community. Of some interests, one might say: a system could scarcely command *sufficient* interest in those subject to its demands for these demands to be acknowledged as obligations, unless it secured to them *this* interest. Thus some claim on human succour, some obligation to abstain from the infliction of physical injury, seem to be necessary features of almost any system of moral demands. Here at least we have types of moral behaviour which are demanded *of* men as men because they are demanded *for* and *by* men as men. Another interest which is fundamental to many types of social relation and social grouping is the interest in not being deceived. In most kinds of social grouping for which there obtains any system of moral demand and claim at all this interest is acknowledged as a claim which any member of the group has on any other; and perhaps most such groupings could scarcely exist without this acknowledgement. When all allowance has been made, then, for the possible diversity of moral systems and the possible diversity of demands within a system, it remains true that the recognition of certain general virtues and obligations will be a logically or humanly necessary feature of almost any conceivable moral system: these will include the abstract virtue of justice, some form of obligation to mutual aid and to mutual abstention from injury and, in some form and in some degree, the virtue of honesty. This guarded recognition of the necessary universal applicability of some relatively vague and abstract moral principles is itself a corrective to the idea of unbounded freedom of choice of such principles on the part of the individual.

I spoke earlier of the need for striking some delicate balances, and I hope that the nature of some of these is now apparent. Constant checks are required if these balances are not to be lost. We have seen in what sense it is true that everyone on whom a moral demand is made must have an interest in morality. But we have also seen that the existence of a system of moral demands (at least as we now understand this concept) requires some degree of general readiness to recognize claims made upon one even when this recognition cannot plausibly be said to be in one's own interest. The existence of some such readiness needs no more to be argued for than the existence of morality in general. But it is necessary to emphasize it in order to correct another exaggeration, the exaggeration which would represent all morality as prudential.[1] To say that this readiness to acknowledge the claims of others does not need to be argued for is not to say that it does not need to be explained. We may discuss its natural sources; and the terms in which we do so will change with the state of our psychological knowledge: the appeal to the concept of sympathy, for example, will scarcely now seem adequate. But, however we explain it, there is no need to sophisticate ourselves into denying altogether the existence or fundamental importance of this recognition of others' claims. Again, we have seen that the fact of acknowledgement of claims may be blown up into the picture of the self-legislating moral agent; and here we should do well to scale down our pretensions to freedom by remembering, if nothing else, the importance of the training we receive and the limited choice we exercise of the moral communities to which we belong. Finally, we have acknowledged some force in the idea of universally applicable principles of moral demand and claim. But to keep within bounds the pretensions of this idea, we must insist again on the flexibility of the concept of a social group, upon the diversity of groups and upon the absurdity of the idea that detailed demands could be shifted indifferently from group to group or apply to all members alike within a group.

There are further important moral phenomena of which the account I have given makes little or no explicit mention. Some of these it might even seem, at first sight, to exclude. Is there not such a thing as moral criticism, from within a society, of the existing moral forms of that society? Cannot different systems of socially sanctioned demand, under which those subject to demands genuinely acknowledge obligations, be the subject of relative moral evaluation? Cannot there be situations in which men may or should recognize moral obligations to each other,

[1] Cf. P. R. Foot, 'Moral Beliefs', *Proceedings of the Aristotelian Society*, 1958–59 (reprinted in P. R. Foot, *Theories of Ethics*, Oxford, 1967).

although there is no common society of which they are members and there is no concept of a 'social' relationship which can be at all plausibly represented as applying to their situation? Any acceptable account of morality must certainly allow an affirmative answer to these questions; and there are others which will suggest themselves. But they no more yield a reason for mistrusting the approach I have adopted than the inadequacy of what I called the minimal interpretation of morality gave a reason for wholly discarding that idea. By enriching the minimal interpretation with certain applications of the notions of interest, and of acknowledgement of obligation, we obtained what was recognizably a concept of social morality. It is necessary only to draw out the significance of certain elements in *this* conception in order to make room for the ideas of moral criticism, and of a morality which transcends standard forms of social relationship. I have remarked already that, because certain human needs and interests are as fundamental and as general as they are, we shall find correspondingly general types of virtue and obligation acknowledged in some form and in some degree in almost any conceivable moral system. Now it is characteristically by analogy with, and extension of, acknowledged forms of these, that moral development proceeds, and that these ideas themselves assume more refined and generous shapes. And moral criticism at its most self-conscious proceeds characteristically by appeal to, and interpretation of, such general moral ideas as those of justice, integrity and humanity: existing institutions, systems of demand and claim, are criticized as unjust, inhumane or corrupt. We may say that so far from excluding the idea of moral criticism, the concept of social morality, as I have outlined it, makes fully intelligible the nature and possibility of such criticism. For we can perceive how the seeds of criticism lie in the morality itself; and we may even hope, on this basis, to achieve some understanding of the complex interrelationships between social and economic change, the critical insights of individual moralists, and the actual course of moral evolution. (It is, for instance, an easy consequence of our principles that moral *formalism* – i.e. a rigid adherence to the letter, with no appeal to the spirit, of the rules – will tend to be at a maximum in a static and isolated society, and that moral *disorientation* will tend to be at a maximum when such a morality is suddenly exposed to radical change.) Just as a social morality contains the seeds of moral criticism, so the two together contain the seeds of a morality transcending standard social relationships. It is easy to see how the tendency of at least one type of self-conscious and critical morality is generalizing and anti-parochial, as it is anti-formalist. Some moralists would maintain that a true concept of morality emerges only at the limit of this generalizing process. This is a judgement in which, as it seems to me, the sense of reality has become quite subordinated to zeal. But wherever we choose to say that 'true morality' begins, I have no doubt whatever that our understanding of the concept of morality in general is best served by the kind of approach that I have sketched. Where what we are dealing with is a developing human institution, it is no reproach to an explanation that it may be described as at least partially genetic.

But now it is time to return to the question of the relation between social moralities and those ideal pictures of forms of life which I spoke of at the outset. All I have so far explicitly said about this is that the realization of any such ideal requires the existence of forms of social grouping or organization which in turn require the existence of a system of socially sanctioned demands on their members. We have since remarked that a system of socially sanctioned demands would fall short of being a system of moral demands unless those demands were not merely enforced as demands, but also at least in some degree generally acknowledged as claims by those subject to them; and it follows from this that to be a member of a moral community cannot merely be a matter of convenience, except perhaps for those who can practise a sustained hypocrisy of which few are in fact capable. Yet it may still be true in general to say that the possibility of the pursuit of an ideal form of life quite pragmatically requires membership of a moral community or of moral communities; for it is extremely unlikely in fact that the minimal social conditions for the pursuit of any ethical ideal which anyone is likely to entertain could in practice be fulfilled except through membership of such communities. But of course the relations between these two things are much more intricate and various than this formulation by itself suggests. The possibilities of collision, absorption and interplay are many. The way I have just expressed the matter perhaps makes most obvious the possibility of collision; and this possibility is worth stressing. It is worth stressing that what one acknowledges or half-acknowledges as obligation may conflict not only, crudely, with interest and, weakly, with inclination but also with ideal aspiration, with the vision that captures the ethical imagination. On the other hand, it may be that a picture of the ideal life is precisely one in which the interests of morality are dominant, are given an ideal, overriding value. To one dominated temporarily or permanently by such a picture the 'consciousness of duty faithfully performed' will appear as the supremely satisfactory state, and being moral not merely as something that matters but as the thing that supremely matters. Or again the ideal picture may be, not that in which the interests of morality in general are dominant, but rather one in which the dominating idea operates powerfully to reinforce some, but not perhaps others, of a system of moral demands. So it is with that ideal picture in which obedience to the command to love one another appears as the supreme value.

This is still to draw too simple a picture. Let us remember the diversity of communities to which we may be said to belong, and the diversity of systems of moral demand which belong to them. To a certain extent, though to an extent which we must not exaggerate, the systems of moral relationships into which we enter are a matter of choice – or at least a matter in which there are alternative possibilities; and different systems of moral demand are variously well or ill adapted to different ideal pictures of life. The ideal picture, moreover, may call for membership not merely of communities in which certain interests are safeguarded by a system of moral demands, but for membership of a community or of a system of relationships in which the system of demands reflects in a positive way the nature of the ideal. For one crude instance of this, we may think again of the morality of a military caste in connection with the ideal of personal honour. In general, in a society as complex as ours, it is obvious that there are different moral environments, different sub-communities within the community, different systems of moral relationships, interlocking indeed and overlapping with one another, but offering some possibilities of choice, some possibilities of adjustment of moral demand and individual aspiration. But here again, at least in our time and place, it is the limits of the direct relevance of each to the other that must finally be stressed. Inside a single political human society one may indeed find different, and perhaps widely different, moral environments, social groupings in which different systems of moral demand are recognized. But if the one grouping is to form part of the wider society, its members must be subject too to a wider system of reciprocal demand, a wider common morality; and the relative significance of the wider common morality will grow in proportion as the sub-groups of the society are closely interlocked, in proportion as each individual is a member of a plurality of sub-groups and in proportion as the society is not rigidly stratified, but allows of relatively free access to, and withdrawal from, its sub-groups. In a political society which thus combines a wide variety of social groupings with complex interlocking and freedom of movement between them the dissociation of idiosyncratic ideal and common moral demand will doubtless tend to be at its maximum. On the other hand an ideal picture of man *may* tend, in fact or in fancy, to demand the status of a comprehensive common morality. Thus Coleridgean or Tolstoyan dreamers may play with the thought of self-enclosed ideal communities in which the system of moral demands shall answer exactly, or as exactly as possible, to an ideal picture of life held in common by all their members. Such fancies are bound to strike many as weak and futile; for the price of preserving the purity of such communities is that of severance from the world at large. More seriously, there may be some attempt to make the whole moral climate of an existing national state reflect some ideal image of human solidarity or religious devotion or military honour. In view of the natural diversity of human ideals – to mention only that – such a state (or its members) will evidently be subject to at least some stresses from which a liberal society is free.

To conclude. I have spoken of those ideal images of life of which one individual may sympathize with many, and desire to see many realized in some degree. I have spoken also of those systems – though the word is too strong – of recognized reciprocal claim that we have on one another as members of human communities, or as terms of human relationships, many of which could scarcely exist or have the character they have but for the existence of such systems of reciprocal claim. I have said something, though too little, of the complex and various relations which may hold between these two things, viz. our conflicting visions of the ends of life and the systems of moral demand which make social living possible. Finally I have glanced at the relations of both to the political societies in which we necessarily live. The field of phenomena over which I have thus loosely ranged is, I think, very much more complex and many-sided than I have been able to suggest; but I have been concerned to suggest something of its complexity. Some implications for moral philosophy I have hinted at in passing, mainly by way of an attempt to correct some typical exaggerations of contemporary theory. But the main practical implications for moral and political philosophy are, I think, that more attention should be concentrated on types of social structure and social relation, and on those complex inter-relationships which I have mentioned as well as others which I have not. For instance, it is hard not to believe that understanding of our secular morality would be enhanced by considering the historical role that religion has played in relation to morality. Or again, I doubt if the nature of morality can be properly understood without some consideration of its relationship to law. It is not merely that the spheres of morality and law are largely overlapping, or that their demands often coincide. It is also that in the way law functions to give cohesiveness to the most important of all social groupings we may find a coarse model of the way in which systems of moral demand function to give cohesiveness to social groupings in general. Similarly, in the complexity of our attitudes towards existing law we may find a model of the complexity of our attitude towards the systems of moral demand which impinge upon us in our social relations at large – or upon others, in theirs.

Finally, I do not think there is any very definite invitation to moral or political commitment implicit in what I have said. But perhaps one question can be raised, and in part answered. What will be the attitude of one who experiences sympathy with a variety of conflicting ideals of life? It seems that he will be most at home in a liberal society, in a society in which there are variant moral environments but in which no ideal endeavours to engross, and determine the character of, the common morality. He will not argue in favour of such a society that it gives the best chance for the truth about life to prevail, for he will not consistently believe that there is such a thing as the truth about life. Nor will he argue in its favour that it has the best chance of producing a harmonious kingdom of ends, for he will not think of ends as necessarily capable of being harmonized. He will simply welcome the ethical diversity which the society makes possible, and in proportion as he values that diversity he will note that he is the natural, though perhaps the sympathetic, enemy of all those whose single intense vision of the ends of life drives them to try to make the requirements of the ideal co-extensive with those of common social morality.

8

Definition of a moral judgement

T. L. S. SPRIGGE

An important distinction between statements of fact and statements of value is widely recognized. Some philosophers are now saying that the distinction has been treated as more determinate than it is, but most philosophers would agree that the distinction is definite and important. The major contributions to Anglo-Saxon moral philosophy of this century have set out to illuminate the nature of this distinction. Ethical statements have been the value statements mainly at issue, but on the whole the aim has not been to show wherein they differ from other value statements, but to show what distinguishes them in common with other value statements from factual statements. The characterizations of ethical statements which have become famous are ones which if they apply to ethical statements at all apply equally to many (or all) other value statements as well.

The question with which I am now concerned is this: What is it that marks off ethical statements from those value statements which are not ethical? In particular I am concerned with the distinction between 'ought' in ethical and in various non-ethical senses.

By an ethical statement I mean a sequence of words which may be said to express a moral judgement. Strictly, the same statement may be ethical in one context and non-ethical in another, but when I call a statement *ethical* or *factual*, or what not, I mean that it is so in its most typical uses.

Reprinted from Philosophy, *1964, pp. 301–22, by permission of the author and* Philosophy.

I start with certain presuppositions not now to be defended. The considerations which make one take up an attitudinist account of value statements are anyway sufficiently familiar. I shall first define what I call an *affective statement*.

I define an affective statement as follows. A statement is affective in a certain context if and only if there is an attitude such that it is a sufficient condition of any one having spoken incorrectly or insincerely that he asserts or assents to that statement in that context without having that attitude. On this definition a value statement may, of course, have factual content.

One might define a purely factual statement in this way. A statement is a purely factual statement in a certain context if and only if there is no attitude such that it is a sufficient condition of any one having spoken incorrectly or insincerely that he asserts or assents to that statement in that context without having that attitude. I shall usually understand 'factual statement' as an abbreviation for 'purely factual statement'. This is of course a quite negative characterization of factual statements. It allows analytic statements as factual, and also any non-empirical synthetic statements, if there are such, which are not affective. (I shall however usually be thinking only of empirical statements when I speak of factual statements.)

To describe someone's attitudes must always be to give some information as to what *does* please or displease him, or what under various circumstances *would* please or displease him. The kinds of attitudes of which there is most cause to speak are likes, dislikes and wants. If I say that someone likes peas I imply that there are quite commonly occurring circumstances under which eating peas pleases him. If I say that someone dislikes French people I imply that he would be displeased at having to share a house with some French people. If I say that someone wants to bring about a revolution I imply that he would be pleased if he thought that he had done something to bring one about. None of these implications are perhaps displayable as entailments. But one may say that unless acceptance of the first set of statements suggested to one acceptability of the second set of statements one would have understood the first set in an odd sense.

I do not say that every statement ascribing an attitude to a man is analysable as a statement about what pleases or displeases or would under certain conditions please or displease him. All I am suggesting is that what distinguishes knowledge about the attitudes of an individual from knowledge about any other aspect of him, is that the former involves knowledge of something (however indeterminate) about what does or under various circumstances would please or displease him.

It is worth pointing out that under my definition 'I love Ermintrude' is never an affective statement. A statement is affective in a given context if and only if there is an attitude, such that it is a sufficient condition of any one having spoken incorrectly or insincerely that he asserts or assents to that statement in that context without having that attitude. Now it may be true that the speaker of 'I love Ermintrude' speaks insincerely or incorrectly if he does not have the attitude, love for Ermintrude, but *I* may assent to his statement correctly and sincerely whatever my attitudes are. I should explain that by 'speaking correctly' I mean 'speaking as a correct user of the language would'. To speak correctly is not necessarily to speak truly. To say someone who makes an affective statement speaks truly is to give one's assent to that statement.

We may now distinguish between those affective statements which are evaluative and those which are not. However, the distinction is one of degree. A statement is evaluative if there is an attitude such that one who asserts or assents to it speaks incorrectly or insincerely unless he has a certain attitude, and wants that attitude shared by others, whereas if it is affective without being evaluative then he speaks sincerely and correctly as long as he has the attitude, whether he wants it shared by others or not. Thus 'This apple-pie is delicious' is merely affective, for one may say it correctly while having no wish that others should like the pie. It would however be a misuse, or an insincere use, of 'beautiful' to say 'This painting is beautiful' if one had no wish that others should share one's delight in the painting.

Emotivists speak as if the main function of value statements is to arouse attitudes. I would not make this a defining characteristic of them. I may use a value statement correctly to convey that I have a certain attitude which I would like shared, without intending my words on that occasion to act as a stimulus for others to adopt that attitude. Doubtless I would like him to agree with me in attitude, and am prepared under suitable circumstances to try bringing this about. The present circumstances may not however be suitable. There is a place for a simple statement of one's views, without any rhetorical purpose. Certainly in so far as my hearer thinks I am likely to have sensible views he will weigh them carefully. But in doing so he need not be influenced by any vibratory power of my language.

These same distinctions apply to terms as well as to statements, in a manner which should be obvious enough.

A term is an affective term (in a certain type of context) if its correct and sincere application (in such a context) to an object indicates that the speaker has a certain attitude to it. Henceforth I shall use 'correct' as an abbreviation for 'correct and sincere'. An argument as to whether an affective term applies to a thing is an indication that the persons involved, assuming that they are speaking correctly, have different attitudes to that thing. To agree on the application of an affective term is to share some attitude.

A term is evaluative if its correct application to something not only indicates that the speaker has a certain attitude to that thing but also indicates that he would like other persons to share that attitude.

Someone might suggest that our definitions should be taken as defining *positive* affective statements and terms, *positive* evaluative statements and terms. In which case negative affective statements, etc., should be defined as the contradictories of these. Our subsequent definitions of ethical statements and terms should in that case likewise be phrased more properly as definitions of positive ethical statements and terms.

A proper discussion of this suggestion is scarcely requisite for present purposes, and might have to be lengthy. One would have to raise such questions as whether 'ought not to' and 'may' are strict contradictories. However, the suggestion may be taken as accepted. The following few remarks may hint at the complicated issues involved.

Consider the statement 'His suicide was cowardly'. This is

affective, for one who affirmed or assented to it and did not feel some dislike for the act in question would be speaking incorrectly. Now consider 'His suicide was not cowardly'. Is there any attitude absence of which would make affirmation of or assent to the statement incorrect (or insincere)? Let us see how one might support an affirmative answer to this question.

It is obvious that one can rule out: 'Yes, some sort of liking for his suicide'. But one might say: 'Yes, absence of dislike'. To this there are two objections. First, it is not required, for one might dislike it on grounds other than those indicated by 'cowardly'. Second, absence of dislike is doubtful as a case of an attitude. Another answer would be: 'Yes, absence of dislike for his suicide as exhibiting fear'. This would still be liable to an objection of the second type.

If the only possible affirmative answers were along these lines I think one would have to allow that the statement: 'His suicide was not cowardly' would not be affective according to our original definition. But if that were taken as a definition of positive affective statement, this would come in as a negative affective statement in the way suggested. For it and 'His suicide was cowardly' would be contradictories in the sense that one could not consistently dissent from both.

But it may be that there is an attitude which is required for correct assent to 'His suicide was not cowardly', namely dislike of anyone else's disliking his suicide as exhibiting fear, or something rather of this sort. In that case, the statement is affective on our original definition. Further, the two statements are not contradictories, for one could consistently dissent from both if one did not oneself dislike his suicide as exhibiting fear, but did not care two pins whether other people did so or not.

There are then two views as to what is involved in saying 'His suicide was not cowardly'. On the one, it is affective on the original definition, but is the contrary rather than the contradictory of 'His suicide was cowardly'. One could reject both. On the other view it is not affective on the original definition, and is the contradictory of 'His suicide was cowardly'. The first view seems the true one of the statement as it would ordinarily be used. If we accept the suggested limitation of our definitions to positive affective statements, etc., and account of negative affective statements, etc., as the contradictories of these we should make it plain that many statements and terms (e.g. 'not cowardly') which would commonly be classed as negative may be positively affective, evaluative or ethical. Indeed the solely negative affective, evaluative, or ethical expression may be rather rare.

It is not the same thing for a term to be emotive as it is to be affective. To say that a term is emotive is to say that it has a disposition to evoke certain emotions under certain circumstances. Affective terms usually are emotive, but may be correctly used under many circumstances in which their emotive disposition is not activated. One uses an affective term incorrectly if one applies it to an object to which *one does not have* a certain attitude. One does not use an emotive term incorrectly because one uses it in circumstances in which it does not *arouse* certain attitudes. It is a sufficient condition of someone having misunderstood my use of an affective term if he does not realize that it indicates *my having* a certain attitude. It is never a sufficient condition of someone having misunderstood my use of an emotive term that *he is not provoked* to a certain attitude.

There is a class of value statements which we may call practical statements. They are statements which express desires or wants (I use these terms as synonyms, and in the widest acceptable sense) that a certain person (perhaps the speaker) or that persons of a certain sort shall act, or under certain circumstances should act, in certain ways. The distinction between practical and nonpractical value statements is one of degree, for an attitude may have a stronger or a weaker relevance to behaviour.

I suspect a tendency among some philosophers to hold that all practical statements (excluding perhaps any which would be affective without being evaluative, a class which anyhow must have few members) which are of any importance are ethical, express moral judgements.

But questions as to what should be done arise continually, and we ordinarily distinguish between those which are moral, or involve moral issues, and those which do not. For instance, political parties sometimes leave their members free on how to vote on a certain issue, because it involves a question of conscience, that is, a moral issue. People debate whether a certain question is or is not moral. Most unilateral nuclear disarmers regard their case as a moral one, but there are those who argue that the question is not a moral but a practical, that is a non-moral practical one. Practical non-moral questions include not only such personal debates as to what restaurant we shall visit, whether we shall play football on a rainy afternoon, but also more impersonal discussions as to whether watching light shows on television is a waste of time, and whether the life of the artist is more worthwhile than that of the business man.

The distinction between categorical and hypothetical imperatives is sometimes supposed to distinguish moral and non-moral value statements. This distinction has various different meanings, but so far as I can see no formulation of it really satisfactorily rules out both non-moral categorical value statements and moral hypothetical value statements. Some aesthetic statements will be categorical on most interpretations. Perhaps it will be replied that moral statements are value statements which are both practical and categorical, and that aesthetic statements are categorical but not practical.

But consider the question 'What career should Jones adopt?' or 'Whom should I marry?' These are practical enough, and in many cases can hardly be called hypothetical. The only desired end which it can be assumed I want is a suitable career, and a good wife, and to find the means to those ends one must decide what would be a suitable career for him or who would be a good wife for me, which is really to ask the categorical non-moral practical question originally intended.

Moreover one can find examples of ethical statements which would be hypothetical on most plausible criteria. Suppose Charles wants to become a concert pianist, but fails to do the necessary practice. I may say to him 'You ought to practise much more'. This may be a moral judgement, a criticism of his lazy failure to use the means to the end he wills. Yet his obligation to play the piano may be hypothetical, inasmuch as it rests on his desire to be a concert pianist. But if one insists that the statement is categorical, because it does more than merely state the fact that a certain act is a means to a desired end, then so is the statement

'You ought to see that show', which is not ordinarily intended ethically.

I wish to find a definition of 'moral' and 'ethical' which shall mark off ethical statements from other value statements, practical or otherwise. Any such definition will be an explication rather than a purely descriptive definition for as ordinarily used the terms are vague. But though the function of any definition we may arrive at is partly that of recommending a certain use for 'ethical' and 'moral', this recommendation will be intended to give a centric rather than an eccentric use of the word. It will aim to make an important distinction, such as it is natural to mark by the words 'moral' and 'ethical' on the one hand, 'non-moral' and 'non-ethical' on the other.

Some have thought that universalizability was the criterion of ethicalness. This concept, like 'categorical', is a complicated and ambiguous one, and I cannot go into it. But any sense of 'universalizable' in which it can plausibly be argued that every ethical statement is universalizable, is likely to mean just about this – 'non-whimsical and non-partial'. In this sense a universalizable value statement is one which expresses an attitude defensible on grounds which one would regard as equally good grounds in any case in which they arose, and on grounds in which references to individuals can be removed in favour of general descriptions. The reference most likely to be irremovable in this way is that made by the first person pronoun. But certainly many value statements which are not moral are universalizable in this sense. For instance, my grounds for saying 'You ought to go on the stage' may be non-whimsical and non-partial, without my piece of advice being a moral judgement. So I do not think that the distinctive feature of a moral judgement is universalizability.

There are cases where a person expresses some sort of attitude and one is in doubt as to whether it is a moral attitude or not. A simple example of this arises when someone says that he is a vegetarian. One often asks in such a case whether he is so on moral grounds or otherwise. If he says that it is because he believes that he in particular or people in general are healthier on a vegetarian diet, his attitude will ordinarily be called a non-moral one. Suppose he says that it disgusts him that animals should be bred merely in order that they may be killed and eaten. Here one will be inclined to call his attitude moral, but even so there will be room for doubt. Does he merely consider it a distaste personal to himself, and allow that those who do not feel any such disgust are morally free to eat meat if they wish, or does he think that those who eat meat – whether they feel disgust or not – are acting in a wrong fashion? Only in the latter case does one consider his view properly a moral one.

In this way it is easy enough to distinguish between vegetarianism held on grounds of personal distaste and vegetarianism held on grounds of moral disapproval. But this 'easiness' arises out of use of the ethical predicate 'wrong'. If, however, to *judge* that eating meat is wrong is to have an attitude unfavourable to eating meat, how does one distinguish such an attitude from the attitude of one who merely feels *disgust* at the idea of eating meat? What is it that in the one case makes it appropriate to use such terms as 'wrong' and 'morally wrong', and in the other case inappropriate? In realizing that moral judgements express attitudes one should not give up drawing the distinction each of us naturally draws between those of his attitudes which are moral and those which are mere tastes.

It is tempting to suggest that ethical statements and terms are those affective statements and terms which are to the greatest degree evaluative. Application of an ethical term to an object indicates that one would like everyone, who considers the matter, to have that attitude, and that one will be disappointed if they do not. Other so-called value terms indicate a milder wish that others should share our attitude; moreover it may be only a certain number of people whose adoption of the attitude we really care about. Thus it may be that the aesthetic critic wishes the class of persons whom he considers cultured to endorse his attitudes, but that unlike the moralist there are many people in whose suffrages he takes no interest.

This view renders it analytic that the only evaluations to which a man would like to convert everyone who considers the matter are his moral evaluations. Although I think it is true that an ethical statement must be highly evaluative in this sense, I do not think all highly evaluative statements are ethical.

A man might hold strong views about the organization of industry, about the principles on which people should choose their careers, about how much money should be spent by governments on the arts, and be interested in winning universal support for those views, and not hold them as properly moral views. He might think that there was nothing actually immoral in action on contrary principles.

Consider the schema 'X ought to do A'. We may distinguish various senses of 'ought' in this schema along two different lines. We can take it that in all cases the speaker, whom we may call 'Smith', indicates his desire that X do A. In the first place we may distinguish cases according to the degree to which Smith wants others to want X to do A. In the second place we may distinguish cases according to the methods which Smith would like to be adopted in order to make or encourage X to do A. These two lines of variation are not unrelated. One could hardly want X to be encouraged to do A by public pressure upon him, without wishing that many other persons should want him to do A.

I have claimed that Smith might be extremely anxious that others should want X to do A without his statement 'X ought to A' being ethical. Suppose we make the schema a sentence thus: 'James Callaghan ought to stand for leadership of the Labour Party'. Smith, talking in January 1963, may want the relevant attitude widely shared, and yet not make his statement as a moral judgement.

Let us now consider the sort of means which I may favour for encouraging X to do A.

I may want him sent to prison if he does not do A. I may simply want certain facts to be drawn to his attention in the hope that it will encourage him to do A. If he does not do A when presented with these facts, I may regret it, but not want any sort of further pressure brought to bear on him.

I suggest that the ethical sense of 'X ought to A' is distinguished from others by the methods which the speaker would favour for inducing X to do A. The statement is ethical if he would like to see X encouraged to do A (if he will not do it otherwise) by the variety of pressures which fall under the term 'operation of the moral sanction'.

Here is the great division. Would we like the moral sanction brought in to back the injunction suggested by an ought-statement? If so it is ethical, otherwise not. The moral sanction is the pressure the group puts upon the individual to act in various ways. It operates mainly by making an offending individual feel an outsider. One who offends against the rules supported by the moral sanction of his society risks cold looks and refusal of ordinary services wherever his offence becomes known.

If I hold that people morally ought to be vegetarians I would like the moral sanction to be directed against those who eat meat. To the small extent that I am myself a representative of the moral sanction I shall be inclined to direct it myself against those who do so. If I do not try to make people who eat meat uncomfortable, that may be for various reasons. I may feel the task hopeless, although if it were possible I would like it performed. I may dislike being eccentric in my relations with people, and although I would like them made uncomfortable about their meat-eating not feel inclined to do anything about it.

In order to gain a clearer concept of the moral sanction let us consider what it is for a moral code to hold in a certain society.

The distinction between a moral code and a legal code is a relatively sophisticated one. The legal code of a society is the system of rules the infringement of which is punished according to a regular form by those who represent the government of the society. A country with no written law may have a legal code in this sense. Statements about an unwritten legal code are hypotheses concerning the actions which the authorities would take under various circumstances. A moral code is not dissimilar. It is a system of rules the infringement of which is punished by signs of unfriendliness towards the offender from average members of the society. He undergoes various distresses stemming from a general withdrawal of liking from him. He may be refused certain services, or he may only suffer through being made to feel disliked.

The moral sanction may take various forms. The form in which we ourselves mainly meet it is as follows.

Human beings have a strong desire to be liked by their fellows. They suffer at signs of withdrawal or absence of this liking.

Partly this is because such a withdrawal may involve the refusal of important services. Partly the consciousness of being disliked, or liked less than one has been, is intrinsically painful. Various words exist in every language which are signs of such a withdrawal of liking, and indicate to some extent the cause of such withdrawal. The term 'selfish' indicates a withdrawal of liking from the person to whom it is applied. It is applied to a man only in virtue of certain sorts of behaviour.

In a civilized society certain rules of conduct are enforced with more or less success merely by the threat of a withdrawal of liking to be evidenced by the use of such words. In a less civilized society this verbal and emotional penalty is perhaps accompanied by a greater withdrawal of services. Where positive physical penalties are used approach is made further towards the legal sanction. A rule is said to be enforced by the moral sanction, when it is adhered to from fear either of the purely emotional penalties or from fear of a withdrawal of services. Of course, those rules which are enforced by the legal sanction, that is by the fear of punishment at the hands of the representatives of the government, are for the most part enforced also by the moral sanction. Various rules, however, are enforced only by the moral sanction.

There is a subtler action of the moral sanction within the individual himself. Often the individual is restrained from behaving in certain ways more by fear of the dislike which one part of him will feel for another part, than by fear of the dislike the group feels for outsiders. This counts as the operation of the moral sanction because he has learnt to dislike himself for acting in certain ways by being made to feel disliked by others for acting in certain ways. The conscience may come to enforce rules different from those from the enforcement of which by others it was at first built up. Where the conscience of an individual enforces rules different from that of his society, we cannot say that the rules are enforced by the moral sanction, if this is still taken as equivalent to the popular sanction. Only where it enforces the rules current in his society can a man's individual conscience be considered an aspect of the moral sanction. But however eccentric its directives may become, the conscience is in the first place the representative of society inside the individual mind.

In order to decide what the moral code of a certain society is, we have to see what actions are penalized by a general withdrawal of liking. Usually adults are deterred from these actions more by the inner sanction, conscience, than by the outer sanction. It is, however, only where the directives of the inner sanction are the same as those of the outer sanction, that its operation is a sign of the moral code in that society.

The moral code is not the only code to be discovered in a society. Apart from the legal code, there are various codes of good form and of aesthetic taste, in particular of dress. If there is a small group in which a particular taste in painting is reprobated, does not this group represent a minority society with a minority moral code? It is perhaps odd to speak of codes of aesthetic taste, but if we do not want the word 'code' to do our job for us we should use it in this rather extended sense.

It is certainly possible for a minority group to have a minority *moral* code. This is something different from a minority group having a minority *aesthetic* code, or code of *etiquette*. The difference seems to be this. A group with its own moral code is to some extent propagandist. It stands in opposition to society at large, and is hostile to the prevailing mores. A group with a minority aesthetic code, and still more with a minority code of etiquette, is content that society at large shall reject its canons. It enforces certain behaviour or tastes on its own members; it does not hope that outsiders will join in the enforcement.

The distinction between a code of morals and a code of etiquette does not turn on the persons to whom it is intended to apply, it turns on the persons by whom it is expected to be enforced. It is quite wrong to say that moral rules, as contrasted with other rules, are rules for us just as being men. In an acceptable sense various rules of medical ethics apply only to doctors. What makes them moral rules is that the doctor who infringes them either is, or those who support the moral rule would like him to be, in disgrace with everyone. On the other hand, one who infringes what is regarded as a rule of etiquette by a minority group, is quite welcome to escape to a circle where no-one will mind the infringement. Thus the distinction between a minority code of etiquette and a minority code of morality is that in the former case the minority group is content to be the only enforcers of the rules, in the latter case it endeavours to make all men its enforcers, or at least regrets that all men are not so.

It will not do, however, to distinguish only between *minority* moral codes and *minority* codes of etiquette. A code of etiquette, of polite behaviour, may be enforced throughout a whole community.

The distinction between a moral code and a code of etiquette is also a somewhat sophisticated one. It can perhaps only be made in a society given to a certain amount of reflection. It would seem to be this. Those who support the moral code of their society would be upset to find that infringement of these rules would be tolerated in certain communities, or would be looked back upon with favour by future members of their own community. Those who support the code of etiquette of their own society are quite prepared that there should be members of other societies quite indifferent to infringement of these rules, or that future members of their society would look back with favour upon infringement of these rules.

This is not to say that a man, A, may not support the moral code of his community and acknowledge that a different moral code is legitimate in a different community. A may grant that the moral obligations of members of another nation or tribe are of a different complexion. What A demands is simply that members of a different nation or tribe should look with dislike upon those members of A's tribe who act against its moral code. He would dislike it if a member of his tribe left it, joined another and were commended for having acted against the moral code of A's tribe. Where rules of etiquette are concerned we may join in an endeavour to enforce them inside our community. We will not object, however, if a member of our tribe is popular among members of another tribe precisely for his deviations from these rules even when in our society.

To say that the moral sanction in a certain society supports a certain rule is to say that people are induced to act in accordance with that rule, partly out of a fear of being rendered outsiders. In some cases there is a threat of actual ostracism. In the last century women who were thought to lead irregular sexual lives might no longer have access to the polite society to which they had previously belonged. If the moral sanction is very successful it will be felt that such a person is rejected by everyone who matters. The fact that there are societies to which a woman could escape where she would not be ostracised is disguised from her by making her feel that acceptance in such a society is of no value as acceptance.

Infringements against a moral code are not all of course punished so extremely. Someone may be made to feel insecure about his place in the group without any actual ostracism.

Members of a certain society may feel that their being accepted as members of that society, with the privileges thereof, depends upon their adhering to a certain degree to certain customs of behaviour. Whether adherence to these customs of behaviour is to be called conforming to a moral code or a code of etiquette depends on how a typical member of that society feels to members of another society who would look without disfavour on members of his society who infringed these customs. If he is quite content that they should do so then they are rather rules of etiquette than of morality. If he objects, then they are rules of morality. The operations of the moral sanction are to be justified before all men.

One might say that a term is ethical if and only if it is incorrect to apply it to an object unless one has some wish to have a certain attitude to that object supported by the moral sanction. Similarly a statement is ethical if and only if it would not be correct to assert it or assent to it unless one had some wish that the moral sanction should be given a particular direction. It is not suggested that the assertion of an ethical statement is always accompanied by a conscious formulation of such a wish, only that one would in fact have been linguistically incorrect to use it if one did not have some such wish, however unconscious. Usually utterance of an ethical statement will be in part an effort to give the moral sanction a particular direction. Always it will indicate a wish that the moral sanction might be given a particular direction. The typically moral *issue* involves a decision on the proper direction of that sanction.

Let us contrast two ought-statements, the one ethical and the other not. Or rather let us contrast two statements, one of which is most likely to be ethical, the other of which is most likely not to be.

(1) 'You ought never to have made love to her. It was a vile thing to take advantage of her loneliness while her husband was away.'

(2) 'You ought not to write such obscure poetry. You ought to produce poetry which would be intelligible to a greater number of people.'

Let us suppose that the 'you' of (1) is Robinson, and the 'you' of (2) is Addington, a poet. The speaker of (1) and (2) is Smith.

It is probable that Smith would like to see the moral sanction impose some discomfort on Robinson. He endeavours to make Robinson regret what he has done, partly by making Robinson feel that he likes him less for what he has done. He hopes too that other people will also be disposed to show Robinson that they feel the same way as does Smith about his behaviour. Smith will be annoyed if Robinson meets people who have no such feelings about his behaviour.

I do not suggest that the only way in which Smith endeavours to make Robinson regret his behaviour is by making him feel disliked for what he has done. He may well try to show Robinson that his act has consequences which Robinson will be sorry about if he thinks of them. The ways in which he may seek to make Robinson regret his act are many and varied. But if it is an ethical statement the attempt or wish to make him feel disliked for what he has done will be present to some degree.

A properly ethical statement indicates that the speaker has a certain attitude which he thinks all decent people will share with him. Decent people are people who feel as he would like all people to feel. Where an ethical statement is intended to arouse regret in some person for something he has done, the regret is intended at least in part to be a regret at being a person who will arouse some dislike in all decent people, that is in all people who feel as the offender would, in his reflective moments, like all people to feel.

When Smith utters (2) he indicates at least this much, that he,

Smith, has been sorry that Addington has written such poetry, and that he would like to see Addington write a rather different sort of poetry. He probably also indicates that he wants other people to feel in similar ways. There are likely to be several aspects of this situation which distinguish it from (1).

In the first place it is probable that Smith is not indicating that he feels dislike for Addington on account of his writing such poetry. He indicates merely, the much less distressing fact from Addington's point of view, that he has wished Addington was writing otherwise. More important, he certainly does not indicate that he would like, nor is it likely that he would like, Addington to be made uncomfortable at the feelings evoked whenever *any-one* hears what sort of poetry Addington writes. He may possibly want Addington to regret having written obscure poetry; he will not want the source of that regret to lie in Addington's sense that he is widely disliked for what he has done. It is likely that Smith would be positively displeased if the penalties of the moral sanction were imposed upon Addington. He would like Addington to be persuaded to change his style, but not by those means.

Smith is not trying to direct widespread public opinion in such a way that Addington feels obliged to comply by his ordinary human wish to be liked and accepted in his society. He would like Addington to feel obliged to comply by other (perhaps more estimable) desires.

It is not impossible that someone should make a moral issue out of the writing of obscure poetry. If he did so he would certainly be wishing to bring the moral sanction to bear in a place where in our society it does not. In Communist countries the writing of certain types of poetry is likely to be branded in a quasi-moral fashion.

T. S. Eliot has written obscure poetry. In so doing he offended some people, but he did not offend against a moral rule of his society. He was not made to feel a widespread withdrawal of liking for having written *The Waste Land*.

There may have been some withdrawal of liking. Many people doubtless felt that he was a charlatan and showed dislike to him on that account. They may have applied words to him calculated to distress him. But these occurrences cannot be described as 'operations of the moral sanction'.

For one thing the withdrawal of liking was not wide enough. Eliot could mix with all sorts of people who evidenced no dislike of him on that account. Nor was this just a matter of a safe coterie of admirers in which he moved. All sorts of people outside the circle of admirers of his poetry doubtless did, and certainly would have, met him, knowing that he wrote obscure poetry such as could give them no pleasure, without any impulse to refuse him services or make him feel unliked on that account. This may well have been true of people such as his doctor, his tailor, and those met at odd gatherings. Although many people may be irritated at obscure poetry, there is no general tendency for obscure poets to be made to feel disliked in our society. What is more, even those people who did vent their irritation had no wish, at least in most cases, that Eliot should be made uncomfortable in every part of society.

If I say 'People ought not to commit adultery' and express thereby a moral judgement I indicate that I like people the less for committing adultery, I want other people to like people the less for committing adultery, and I want people to be discouraged from committing adultery for fear of the penalties of the moral sanction.

If we enquire further as to the nature of these penalties we may find something puzzling. These penalties fall into two classes, refusal of services and indications of dislike. Yet the more the penalties are those of the former class the nearer the moral approaches the legal sanction, and sets up a sort of unofficial analogue to positive law. It is in the latter class that the typical penalties of the moral sanction fall.

The main form which these indications of dislike take is the public application of condemnatory ethical terms to the offender. This is odd, for it is the function of ethical terms that we are investigating. We seem to be saying that ethical terms are used mainly to indicate our wish that other people shall cause pain by the application of ethical terms. Do we mean that in calling X wrong, we indicate our wish that others shall call X wrong, thereby indicating that each of them wants all the others to call it wrong, and so on?

A simplified answer would be this. For the moment let us confine ourselves to condemnatory ethical terms. There is an expressive and a reflective use of ethical terms. Used expressively an ethical term simply evinces dislike in much the same way as does a snarl or a curl of the lip. This expressive use is highly emotive. For reasons both simple and complex a person is pained by any sign of dislike on the part of another. Therefore to use an ethical term expressively in the presence of the person to whom it is applied is to give him pain. There is also the reflective use. Here the term serves not as a primitive and pain-giving evincement of displeasure. It indicates, perhaps in perfect coolness, that the speaker has certain dislikes and would like others to have them too, and would like these dislikes to be evinced by ethical terms used expressively. It indicates further that one would like the thing one dislikes checked by fear of the pain given by expressively used ethical terms.

This is simplified because it is obvious that no clear line exists between reflective and expressive use of ethical terms. One might clear this up by saying that these were two extremes, and that the use of an ethical term might fall at any point between them.

There is, however, a difficulty in the idea of a purely expressive ethical term. If it does no more than merely evince dislike, in what sense is it ethical? How does one apply all our toil on the distinction between ethical and other affective terms to it? Is it ethical because it evinces dislike in a situation which might suit the wants indicated by someone's reflective use of an ethical term? It would not perhaps be impossible to find a criterion of 'ethical' applicable to such expressive uses. We need not seek it, however, as a better way of putting the matter suggests itself.

We may stick to our former definition of 'ethical term'. In every application of an ethical term one indicates the presence of three features, an attitude of one's own, a wish to have that attitude shared, a wish to have it supported by the moral sanction. These three features are present, however, in very various degrees, and even in different ways. In the most reflective use of the term one deliberately lets one's listener know that one has these three attitudes (the primitive attitude and the two wishes about it). In the most expressive use of the term one gives vent to the primitive attitude spontaneously, or, what is hardly distinguishable, seeks to give him pain by using words ordinarily indicative of such an attitude. In this case there is no deliberate wish to indicate the presence of the two second-order attitudes. What makes it the use of an ethical term is, however, the fact that one does have these second-order attitudes. On enquiry someone could find out that one is keen that others should share the attitude evinced and give vent to it in modes typical of the operation of the moral sanction.

Our account leaves room for all degrees of spontaneity and deliberateness. The second-order attitudes may be remote from consciousness, existing only dispositionally, or they may be carefully formulated principles of action. The first-order attitude itself may be at the moment largely inactive, and the speaker may be merely letting his listener know that it is there to be called into action on occasion, or it may be spilling over into the ethical language spontaneously. Moreover the ethical term may have been used in a deliberately emotive fashion, in order to give pain or pleasure, or, at another level, in order to produce a shared attitude. Alternatively it may have an emotive force which has arisen without calculation either from the listener's distress or pleasure at the emotion indicated spontaneously or deliberately by the speaker, or from his tendency to adopt the attitudes of that speaker.

The penalties and rewards of the moral sanction are the penalties and rewards received at the hands of public opinion. The opinion in question is public in two senses. Firstly it is the opinion of nearly every member of the society, or where it is singular judgements that are in question, it is the opinion which almost every member who considered the matter would have. Secondly it is an impersonal sort of opinion, not felt as a mere private taste, but as the attitude of every decent person. What one who holds an ethical view would like to see is support for his attitude from as many persons as possible from every society leading to a universally consistent operation of the moral sanction. It leads to this if and only if every person who considers action in dis-accordance with this view is faced with the prospect of social discomfort in so doing.

This account does not put any logical limitations upon the content of the moral law. When a man speaks of the moral law he speaks of a system of evaluations and rules which he would like to see enforced by the weight of public opinion, in particular by the threat of a general withdrawal of liking from those who infringe it. But these evaluations and rules may logically have any content whatsoever. The moral neutrality appropriate to metaethics is retained. It is not said that one *could* not think it morally obligatory to clap the hands three times in an hour.

On the other hand it is a richer account of moral language than is given by one who says no more than that it is affective, or that it is emotive, or that it is prescriptive. It distinguishes moral views from other evaluations, without doing so by way of their primary content.

It has sometimes been suggested that ethical questions are distinguished from other practical questions by being the practical questions which arise when two or more individuals or groups clash. Ethical statements and principles are proposals for resolving, regulating and preventing such clashes.

The present theory gives some explanation of the plausibility of this account, without granting its complete validity. That the moral sanction should, and should be desired to, attach itself mainly to rules governing relations between one person and another, rather than to problems affecting mainly the interests of various single individuals, is easily understandable. The moral sanction exists, and is wanted to exist, because a lot of people are interested in seeing a check put upon certain kinds of conduct. It is in most people's interest that the violence and greed of others should be controlled. Each of us has much less interest in seeing the more purely personal choices of others controlled in this way; nor, for the most part, do we wish to see an increase in the control others have over our own personal choices.

These being the interests of each of us, it is to be expected that the moral sanction should attach itself, to the general satisfaction, mainly to those aspects of a man's conduct which interfere with the interests of others. These are the aspects which some philosophers hold it a tautology to say are alone proper objects of moral judgement.

There are, however, forces working in the other direction, forces which make us want others to have more control over us than is in their obvious interests, and us to have more control over others than is in anyone's obvious interest. If interest be used in a wide sense, so that every sort of desire represents an interest, then we may say that there are subtler interests at work than the mere desire for physical safety and comfort.

Hatred of the odd in others leads us to extend the area of social control beyond what is needed to preserve us from straightforward harm. Hatred of the odd in ourselves leads us to demand that society should keep stern order.

There is no choice so private that it is logically absurd to treat it as a moral choice. There is no sphere of life on which a man cannot with consistency seek to bring the moral sanction to bear. It is not a tautology that moral issues always concern *clashes* between individuals or groups.

Those who would emphasize the inward aspects of the moral life may find this emphasis on social control offensive, however open the content of the Moral Law may have been left.

Suppose Robinson's brother has committed a vicious crime which Robinson regards as truly reprehensible, and he asks for Robinson's assistance in avoiding the legal penalty. This might pose what should obviously be called a moral problem. One could not compare it to a choice like that one has to make of a career, and say it is a practical choice, not a moral choice. Nor could one compare it with the choice of what party to support, which often involves decisions of value without involving moral decision. It is the very type of a difficult moral problem. Nor is it a struggle between duty and interest. It is, or at least it may be, a struggle between two moral assessments of the situation. It may be argued that Robinson would be a pretty poor fish if his main concern in this struggle was as to which course society would blame, which praise. There are all sorts of features in the situation which might concern him. Probably the operation of the moral sanction would play little part in his thoughts, certainly not the determining part.

It must be remembered that on my account such a question would be moral if it involved the question what society should demand of him. This is different from the question what society does demand from him.

Even so, it may be thought that my account suggests that moral struggle is always a struggle as to how one may please an imaginary set of onlookers, who since they are supposed to speak with the voice of society (even if ideal society) will speak with all the crudeness typical of public opinion.

Such a description describes very well some cases of moral struggle, but I agree that it does not do so for all, and it may not do so for the one above. Robinson's thoughts may be of a different cast. He may consider all that he owes to his brother on the one hand, and the depravity of his crime on the other. He may well resolve his problem without paying any attention to the question how society should encourage and discourage certain activities.

This is not denied. No limitation is put upon the manner in which a moral issue may be determined. What is said is that anyone considering Robinson's struggle, or any other struggle, may correctly call it moral if he considers that society should concern itself with how the struggle is resolved. If he thinks it desirable that social discomfort of a certain sort should follow on Robinson's deciding the matter one way then he thinks that Robinson's decision raises moral issues.

What of Robinson? He may correctly speak of himself as faced with a moral issue if he confesses that society should interest itself in how he resolves it. He does not have to make any such profession in point of fact. It is enough to render it correct for him to think of his issue as moral, if, when the point arises verbally or in practice, he admits that it is a matter in which society should take an interest. If, on the other hand, a man is struggling over something, such as what career to adopt, and thinks that it is no business of society's to penalize him for choosing in a certain way, then he should not call his struggle moral.

Society does, I believe, have some introspectible presence in most moral struggles. For in a moral struggle a man is inclined to think of himself as bound to justify himself to any man who, in appropriate circumstances, queries his decision. He does not have to feel that he would persuade *any* man that he acted well, only that he would persuade any decent man. He feels that if he did not produce a justification which would persuade any decent man, it is within such a man's rights to give him pain by signs of his displeasure.

To say that a man is involved in a moral struggle may indicate either of two things. It may indicate that the speaker considers it a struggle in which society should concern itself, or it may indicate that the struggler would allow that society's concern was appropriate. That is, we may distinguish between saying that a man is faced with what actually are moral issues, and saying that a man is faced with what he takes to be a moral issue.

I wish now to consider whether my account of 'ethical' and 'moral' struggles in an *ethical* view illegitimately. I should say that it did not, for it leaves the content of morality completely open. Here it differs from various other metaethical theories.

Some philosophers have denied that a consistent ethical racialism was possible. They have thought that on purely meta-ethical grounds one could deny the possibility of a genuine moral judgement to the effect that different races had different rights. Professor Duncan-Jones in *The Moral Philosophy of Bishop Butler*, page 173, has given an account 'of the term "moral judgement" in such a way as to exclude some attitudes to which the name "moral" is sometimes given. We have excluded what is sometimes called "tribal morality": that is, if a man favours a given type of action when it is done by a member of a certain class, or totem, or a person with a skin of a certain colour, but condemns actions of the same type when they are done by someone of a different class, totem, or skin-colour – however consistent he may be in the application of his tribal rules – his attitude will not fulfil the conditions of universalizability' – and therefore will not constitute a moral judgement. The only case in which moral judgement may take account of racial distinctions is one in which they affect a man's abilities with regard to the performance of a favoured or disfavoured act.

On the present account racialism is not excluded on this *a priori* ground.

No one can doubt, for he knows that it occurs, that a racialist code may be enforced within a white community, constituting part of a larger multi-racial community. The white member of such a community who disobeys the racialist rules meets punishment from the moral sanction. He may find himself generally ostracized. There may also be legal penalties, but these do not concern us at present. Moreover it is not out of the question that within a multi-racial community a racialist ethic may be supported by public opinion both of black and white. Certainly a man may be in favour of enforcement of such rules by both sides.

In a society in which there is a dominant race, and another race which is discriminated against, the dominant race may enforce racialist rules on its own members by threatening social penalties for their infringement. When we consider the attitude of the dominant race to the dominated race we see that there are two possibilities. Its members may hope and expect, or at least wish, that the members of the dominated race will themselves play their part in enforcing the code by their own public opinion. Alternatively they may concern themselves not at all with the attitudes of the dominated.

Possibly many whites in South Africa support a code of race-discriminatory rules, which they would like to see supported by the public opinion of both white and black. They may want blacks as well as whites to be shocked by racial fraternizing, and want them to penalize by their own disapproval blacks, and even whites, who infringe the rules. If one considers anti-semitism in Nazi Germany one may perhaps find the other alternative realized. It is doubtful whether the Nazis can have had any wish, let alone hope, that their anti-Jewish policy should have the moral support of the Jews themselves. They can have hardly hoped or wished that the Jews should socially penalize each other for not co-operating with the persecution.

This is an important difference. To favour a code regulating the relations between two races, and want to win moral support for this code from both sides, is one thing. To favour a code regulating the relations between two races, and have no interest in the attitude to it of one of these races, is another thing. On our account the first would be a moral code, the second would not.

It is a characteristic of an ethical statement that one who accepts it feels it incumbent upon him to justify the statement to all men. He may expect that he will in fact fail to gain many suffrages, but at least he wants to do so. If he formulates ought-statements, and is quite indifferent as to how many people affected by action upon them will regard them, then he gives up the attempt to take a moral line at all.

This does not put any logical limitation upon the content of a moral code. It does explain why, diverse as it is, moral opinion generally falls within certain limits. We do not, on the whole, want something when we have no hope of it. People usually have enough sense not to hope for support for a system of rules from those who can clearly see that it conflicts with their major interests. This puts some check upon the varieties of moral opinion.

A moral code is a system of evaluations and prescriptions conformity to which is encouraged in a society by public opinion. It differs from a code of etiquette in that its supporters are concerned that public opinion in all times and places shall back it up or regard it with sympathy. Its supporter may allow that a different moral code would be suitable for a different society, but he must hope that members even of that society would sympathize with enforcement of his society's moral code in his society. He may make the same point by regarding his and all other acceptable moral codes as special applications of a wider moral code which he would like to see enforced by public opinion everywhere on everyone. Public opinion backs up a moral code mainly by the threat that those who infringe it will be less readily accepted members of any community where that public opinion operates. To some extent it may also offer rewards, for it may offer the man who goes beyond what is required to avoid condemnation the prospect of being universally honoured and loved.

A statement is ethical in a given context if and only if there is some attitude such that the person who asserted it or anyone who assented to it in that context will have spoken incorrectly or insincerely, one might say, will have spoken misleadingly, if he does not have that attitude and want it widely shared and enforced (if necessary) by the moral sanction, that is by group pressure upon the individual. But though ethical statements, sincerely and correctly uttered, are always indications of an attitude to the operations of the moral sanction, it is not even generally true that to indicate this is the main purpose of their utterance. What is given here may serve as a defining characteristic, a characteristic shared by all and only ethical statements. It is not the most interesting or important characteristic of most subclasses of ethical statements.

There are many questions of value which are not moral questions. Many of these are more interesting, more perplexing and in some ways more important than most moral questions. It may even be desirable to extend the area of non-moral value at the expense of morality.

Morality is something with which we should conform. The projects we set ourselves must not conflict with it. Yet we would be fools to seek the point of life in morality. Morality exists for men who have something worth doing other than being moral. It is the limitation society properly imposes upon the individual's pursuit of his own ideals.

Moral issues are everyone's concern. Moral issues are issues with which we are concerned that everyone should be concerned. Moral views are views which we wish everyone to adopt and enforce by the giving and withholding of love. But our moral life is not the most important aspect of our life. We should conform to morality but seek our satisfaction elsewhere.

9

The concept of morality

W. K. FRANKENA

I

When I was young, teachers had an exasperating way of asking, 'What is a so-and-so?' and then scolding when we answered by saying, 'A so-and-so is when . . .' Nevertheless – perhaps in a belated burst of rebellion – I mean to answer the question. 'What is Morality?' by saying roughly, 'A Morality is when . . .' It might even be better to say that the question I propose to consider is not 'What is Morality?', but 'When is Morality?' There are so many dimensions and forms in which one can ask and answer the question 'What is Morality?' that one can find a manageable subject for a lecture such as this, and achieve clarity and direction in dealing with it, only by selecting one of the many questions suggested by that larger question. However, the question I have selected, 'When is Morality?' may seem even less clear than the original one. I remember hearing an older acquaintance belaboring an army intelligence test for its stupidity. 'What do you think they asked me? They asked me, "When is a fish?" How was I supposed to know the answer to that one?' I am not sure I know the answer to my own question, but I can at least explain what I mean by it.

I mean, of course, to ask 'When are we to say of something that it is moral or a morality?' or 'What are we to take as the necessary and sufficient conditions of something's being or being called

---

Reprinted from University of Colorado Studies, *Series in Philosophy No. 3* (*1967*), *pp. 1–22, by permission of the author and* University of Colorado Studies.

moral or a morality?' But even when put in this form my question needs some explanation and hedging around. (1) The question is not, 'When do or may we say that a man or society is living rightly, is morally good, or is enjoying the good life?' An individual or group may have a morality without being morally good, living rightly, or having a good life, even by his or its own standards. This is, indeed, all too often the case. It is not 'right' or 'good' we are trying to define, but 'moral' and 'morality'. (2) These terms, however, are ambiguous. We often use 'moral' or 'ethical' as a synonym for 'morally right' or 'morally good'; in fact, it is almost a vogue to say 'That wouldn't be ethical' or 'It isn't the moral thing to do' instead of using the older-fashioned words 'right' or 'good'. Here 'moral' is used as the opposite of 'immoral', and it follows from what was just said, that we are not asking for a definition of 'moral' in this sense. Similarly, 'morality' is sometimes used as equivalent to 'rightness' or 'moral goodness', or perhaps oftener as equivalent to 'the moral quality of an act or person', whether this is plus or minus. Here 'morality' is used either as the opposite of 'immorality' or as covering both 'moralness' and 'immoralness' (if I may use these terms), much as 'temperature' is used to cover heat and cold. Philosophers have often asked after the criterion or definition of morality in this sense, i.e. for the distinction between right and wrong or between goodness and badness. But, again, this is not our question here. For 'moral' is also often used, not as the opposite of 'immoral', but as the opposite of 'non-moral'. Thus, we speak of moral rightness and of other or non-moral rightness, of moral duties and of legal ones, of moral judgements and of non-moral ones, of moral questions and other kinds of questions, of moral sanctions and natural or religious ones, of moral considerations and aesthetic ones, etc. It is about the definition of 'moral' in such uses, i.e. as the opposite of 'non-moral', that we are asking now. So too, we often speak of morality as a kind of activity or institution which is different from but may stand in relations with law, religion, prudence, convention, art, science, etc., and which involves a certain point of view, the making of certain sorts of judgments, reasoning in certain ways, employing certain kinds of sanctions, etc., etc. Our question is about morality in this sense. What are the criteria for distinguishing morality and what pertains to morality from other things, things that are non-moral but not necessarily immoral? (3) However, even when they are used to mark off the moral from the non-moral rather than the immoral, the words 'moral' and 'morality' are used in at least two different, if related, ways. First, they are used in such expressions as these: 'Have the Navaho a morality?', 'What is the ethics of the Hindus?', 'His moral principles are nothing to write home about', 'The Christian morality is different from that of the Chinese', etc. Here morality and moral principles are thought of as belonging to some individual or group, and as varying from individual to individual and from group to group. Thus, we speak of 'moralities', 'a morality', 'his moral principles', etc. Second, we also speak of 'the moral law', 'the principles of morality', 'the moral point of view', and so on, and here we seem to mean something that is not relative and does not belong to any person or group. I shall be concerned in this lecture with the use of 'moral' and 'morality' in expressions of the first sort; although I shall keep expressions of the second sort in mind (because their analysis has a bearing on the analysis of those of the first), I shall leave that analysis for another occasion. In short, I shall ask only, 'When are we to say of some individual or society that he or it has moral principles or ideals, a moral code, a morality?'

Yet two more of these preliminary explanations and qualifications of our question. (4) In asking, 'What is it for *x* to have a morality?', one might put the emphasis on the 'to have', and ask questions like this: 'Must one have a habit of acting on a certain moral principle before he can be said to have it?' This is not our problem. We shall be concerned, not with the question 'What is it to have a morality?', but with the question, 'When one has a morality, *what* is it one has?' – 'What is it for a code to be a moral code?' (5) Philosophers have sometimes asked lately if a morality must consist of (or at least contain) rules or principles of duty and obligation, or if it may consist wholly of ideals of virtue or moral goodness. This is an interesting and important question, but it too must be left for another occasion. In what follows I shall try to allow for both possibilities –

hence my frequent use of the phrase 'principles or ideals'.

We must also note that our question and our answer to it may take two forms. For when we ask what morality is or what is to be regarded as built into the concept of morality, we may be asking what our ordinary concept of it is or entails, what we actually mean by 'moral' and 'morality' in their relevant uses, or what the prevailing rules are for the use of these terms. And here the question is not just what we do say, for what we do say may not conform to our rules or it may be less than our rules permit us to say. It is our rules for using 'moral' and 'morality' that we want, not our practice – in Ryle's terms, our use, not our usage – what we may say, not what we do say. However, when one asks what morality is or how it is to be conceived, one may be interested, not so much in our actual concept or linguistic rules, as in proposing a way of conceiving it or a set of rules for talking about it, not so much in what our concept and uses are, as in what they should be. If the questions are taken in the first way, the discussion will be a descriptive-elucidatory one, and the arguments pro and con will have a corresponding character; if they are taken in the second sense, the inquiry will be normative, and the arguments will have a different character, though, of course, one may still take the fact that we actually think and talk in a certain way as an argument for continuing to do so.

Now, most recent philosophers who have dealt with our topic have been shy about making proposals of a normative sort, not only about how we should act but even about how we should think and speak. Though some of them do at least favor one way of speaking as against another, they tend to try to rest wholly on the basis of actual use and its rules. Indeed, they have tended to think that philosophers as such should not venture to propose revisions of our moral concepts, since to do so is to make a normative or value judgement, possibly even a moral one, and the business of philosophy is or should be (a normative judgement!) 'analysis' or 'logic'. It appears to me, however, that they have in fact been normative at least to the extent of implying that we ought to go on talking and thinking in accordance with the prevailing rules, only more faithfully. But, if one may or must be normative at all, then in principle there is no reason why one may not be revisionary, especially if one finds difficulties and puzzles in our ordinary manners of thought and expression. In what follows, at any rate, I shall take it to be appropriate for a philosopher to ask whether something should be built into our concept of morality, even if it is not. In fact, though I shall pay a good bit of attention to our actual concept or concepts and to our actual uses, and rest on them when I can, I shall take our problem to be primarily a normative rather than a descriptive-elucidatory one.

In doing this I am particularly pleased to have the support, not only of R. B. Brandt, but of G. H. von Wright. Describing the nature of his 'conceptual investigations' in *The Varieties of Goodness*, von Wright writes:

> An urge to do conceptual investigations . . . is *bewilderment* concerning the meaning of some words. . . . We are challenged to *reflect* on the grounds [for calling things by a certain word] . . . when the grounds have not been fixed, when there is no settled opinion as to what the grounds are. The concept still remains to be *moulded* and therewith its logical connections with other concepts to be *established*. The words and expressions, the use of which bewilder the philosopher, are so to speak *in search of a meaning*.[1]

The concept of morality and the words 'moral' and 'morality' seem to me to be eminently cases in point.

Unlike von Wright, however, I am still willing to call my inquiry meta-ethics, though I am not so sure about the distinction between meta- and normative ethics as I once was. Even if meta-ethics is thought of as normative in the way indicated, as making recommendations about the criteria by which we are to apply such words as 'moral' and 'morality', it does not follow that it is being converted (or subverted) into normative ethics. For normative judgements are not necessarily moral judgements (except on a wide, loose usage which is unfortunately becoming too current among philosophers). Whether a normative judgement is moral or not depends on the kind of reason that is used to support it, and the reasons given for a proposal about the application of 'moral' need not be moral. They may consist of such facts as that the proposal accords with ordinary use, that it

[1] pp. 4–5.

eliminates a certain ambiguity or vagueness, or that it makes for a clear way of distinguishing morality from law and for relating the two to one another.

II

We may begin with a point about morality on which all contemporary moral philosophers seem to be agreed, namely, that morality is and should be conceived as something 'practical' in Aristotle's sense, i.e. as an activity, enterprise, institution, or system – all of these words are used and it is hard to know which is best – whose aim is not just to know, explain, or understand, but to guide and influence action, to regulate what people do or try to become or at least what oneself does or tries to be. Two typical quotations will serve to present this point:[1]

> Moral discourse is a practical, directive mode of discourse functioning to guide conduct and alter behavior.
>
> . . . this much at least can be said about an ethical system or way of life, that its function is to guide conduct . . . the notion of such a function is built into it . . .

If this is true, then we may say that *x* has a morality only if he has something – a set of rules, principles, ideals, etc. – that he takes as a guide to action. There are, however, other systems besides morality that have a built-in function of regulating conduct – law, convention, prudence, and religion. The question, then, is this: when is an action-guide (an AG) to be considered a morality? When should we regard a mode of regulation – a set of maxims, principles, ends, or ideals – as a moral one?

This question has been much discussed of late, and a variety of views have been suggested in answer to it, either in the form of descriptive elucidations of our actual thinking or in the form of normative proposals. I cannot deal with all of these views or with all of the problems involved in our question, however, and therefore propose to centre on one debate that is going on in recent moral philosophy, namely, that between those who favor a 'narrower' concept of morality as having a 'material' *social*

[1] Cf. Kai Nielsen, 'Wanton Reason', *Philosophical Studies*, XII, 1963, p. 71; J. Kemp, *Reason, Action and Morality*, 1964, p. 191.

feature built into it and those who stand for a 'wider' concept of morality as having only certain 'formal' features built in. The one view maintains that some kind of sociality should be regarded as a necessary condition of an AG's being a morality, and that other conditions are not sufficient; the other insists that other conditions should be regarded as sufficient and that sociality is not necessary. My discussion will thus be a continuation of my paper in the volume edited by H. N. Castañeda and G. Nakhnikian,[1] and it will take the form of a defense of the narrower concept of morality. Most recent philosophers seem to be on the other side, though the positions of some of them are unclear on the question at issue, but I can claim some allies, e.g. S. E. Toulmin, Kurt Baier and J. Kemp.

Even this debate about 'the social definition of morality' I cannot deal with exhaustively. I shall proceed mainly by way of reviewing the two recent discussions that seem to me most interesting and stimulating, viz., those of R. M. Hare in *Freedom and Reason* and W. D. Falk in the volume just mentioned, though I shall also make some reference to other writers.

III

If our actual use of the terms 'moral' and 'morality' were clearly in favor of either the narrower or the wider concept of morality, I would regard this as decisive, for if we always used them in one of the two ways (except for clearly extended or parasitical uses), there would be no point in advocating that we use them in the other way. In fact, proponents of the wider use do not deny that there is a narrower use, though they do sometimes ignore its existence. Some of them, like H. L. A. Hart, think that ordinary discourse favors the wider concept, but Falk at least seems to think it favors the narrower one. Advocates of the narrower concept, on the other hand, are apt to assume that there is no wider one or to regard it as eccentric. I should, of course, like to join them in this position. Actually, I am not entirely convinced that there is a clear, correct use of 'moral' and 'morality' in the wider sense. When Butler says that the Epicurean system of

[1] *Morality and the Language of Conduct*, 1963.

conduct is not the 'moral institution of life', and when Kant distinguishes between 'counsels of prudence' and 'commands of morality', they ring a bell in my mind. We do call the Epicurean AG an 'ethics', but would we call it a 'morality'? When we speak of the morality of the Epicureans, we refer to their actual conduct, not to their views about what conduct should be like. There is a recent book with the title, *The Morality of Self-Interest*;[1] but it is not really claiming that a purely egoistic AG is a morality; it maintains, in fact, that actions are not right in the moral sense unless they promote social as well as individual welfare. John Ladd has contended that the AG of the Navaho is a kind of egoism, but it is interesting that he calls it an 'ethics' rather than a 'morality'.[2] We are more liberal with the term 'ethics'; it serves not only as a substitute for 'morality' but also as a label for the normative theories proposed by philosophers and other intellectuals. We say, e.g. 'the ethics of immoralism', but not 'the morality of immoralism'.

I am, however, sufficiently swayed by the contentions of those who oppose building any kind of sociality into the definition of morality to be willing to admit that there may be such a wider sense of 'morality' as they describe. Karl Jaspers' usage does not seem to me entirely unnatural when he says:

> . . . [Nietzsche's] attack on morality is not an attack on morality in general, but an attack on one morality through another morality. . . . Where we put an unconditional value on something, there we speak morally; and vice versa, where we speak morally, there we are taking something as unconditional . . .[3]

Neither does it seem wholly forced when eighteenth-century writers call egoism a 'scheme of morality'. In fact, I am inclined to think that the following situation both shows that there is a narrower (social) sense of 'moral' and at least suggests that there is a wider one (such as Jaspers has in mind). We sometimes think that the requirements of morality and those of prudence or self-interest may conflict. Now suppose we ask, 'When they conflict, what should one do? Which takes precedence, morality

[1] R. G. Olson, *The Morality of Self-Interest*, 1965.
[2] *The Structure of a Moral Code*, 1957.
[3] K. Jaspers, *Nietzsche*, 1936, p. 124 (my translation).

or prudence?' It then seems natural to say that this is a moral question. But, if we say this, then it appears that morality comes in in two places, once as one of the rival claimants, and again as the final court of appeal that judges between them.

I shall, therefore, assume that there is, as Falk maintains, a semantic case for saying that there are in our actual discourse two uses of 'moral' and 'morality' corresponding to the two concepts involved in our debate. That, I suggest, is why Nietzsche shifts ambivalently between thinking of his 'master' AG as beyond morality and thinking of it as a new and higher morality. As Tillich puts it:

> 'Transmoral' can mean the re-establishment of morality from a point above morality, or it can mean the destruction of morality from a point below morality.[1]

Our question, then, is whether we should go on using 'moral' and 'morality' in both senses, use them only in the wider sense, or use them only in the narrower one. I take Falk and Hare (and Hart) to be favoring the second alternative, and mean myself to favor the third.

Of course, there is a fourth possibility – to give up the use of the words 'moral' and 'morality'. Sometimes Hare seems on the verge of suggesting this. However, to advocate this move is to propose a rather drastic revision of our vocabulary. To quote Tillich again, but reading 'morality' where he says 'conscience',

> The famous theologian, Richard Rothe . . . has made the suggestion that the word ['morality'] should be excluded from all scientific treatment of ethics, since its connotations are so manifold and contradictory that the term cannot be saved for a useful definition. If we look not only at the popular use of the word with its complete lack of clarity, but also at its confused history, this desperate advice is understandable. But . . . it should not be followed, for the word ['morality'] points to a definite reality which, in spite of its complexity, can and must be described adequately; and the history of the idea of [morality], in spite of the bewildering variety of interpretations that it has produced, shows some clear types and definite trends.[2]

Both Hare and Falk remark that the question how to use the

[1] P. Tillich, *The Protestant Era*, Phoenix Books, 1957, p. 148. [2] Ibid., p. 136.

words 'moral' and 'morality' is 'only a terminological question, albeit one to which some answers can gravely mislead'.[1] What matters, they insist, is to be clear what the substantive questions are, and to see that no substantive questions are settled by any terminological legislation. In fact, both of them then refrain from offering a ruling on the use of those words, though they favour using them only in the wider way. I agree that the question is a terminological or verbal one, that some answers to it are apt to be misleading, that the important thing is to be clear about substantive matters, and that linguistic decisions cannot settle any substantive question. But, unless we drop the words in question (itself a linguistic decision), we must have some ruling about their use – which concept they are to stand for – even if it is only to the effect that we should go on using them as we have. I do not really mean a 'ruling', of course, for one can give a ruling only if one is in a position of authority of a certain kind, and I do not believe that philosophers as philosophers are in such a position; I mean a proposal or recommendation, and I do believe that philosophers as philosophers may and should make terminological proposals or recommendations, provided they give reasons for making them.

So far I have described only very roughly the two views of morality that I am opposing to one another, and it will be helpful if I now state them more accurately. The view of my opponents may be characterized as asserting that:

> x has a morality or moral AG if and only if he has an AG of which such conditions as the following hold:
> (a) x takes it as prescriptive,
> (b) x universalizes it, wills it to be a universal law,
> (c) x regards it as definitive, final, overriding, or supremely authoritative or supremely important.

They differ about which of these features are necessary and sufficient. Jaspers in the above quotation seems to hold that (c) is necessary and sufficient, but probably he means to include (a) in (c). I take this also to be the position favored, though not actually

[1] Hare, *Freedom and Reason*, 1963, p. 147; cf. Falk, 'Morality, Self and Others', in *Morality and the Language of Conduct*, ed. by H. N. Castañeda and G. Nakhnikian, 1963, pp. 30f., 33, 53.

affirmed, by Falk. Hare, as I understand him, asserts that the three features are all necessary and jointly sufficient.[1]

As for my own view – there are many senses in which morality may be and has been held to be essentially social, but I am not concerned to defend all of them on this occasion. In some of them, some of my opponents are willing to say that morality is social, e.g. in insisting that moral judgements must be universalized Hare is ascribing one kind of sociality to them. Indeed, it is only a minimal kind of sociality that I want to defend now, though it involves more than Hare is willing to admit. I am not now interested in arguing that morality is essentially a social institution or organ, though I am inclined to think so. Nor am I contending that it has a social function of maintaining a stable society or promoting the general welfare; I am not proposing to build utilitarianism into the definition of morality. I have no wish even to claim that a judgement or rule is moral only if it involves the relation of one individual to others (and not merely to himself) though, again, I am inclined to believe this. Just now I am willing to allow that a moral AG may include self-regarding duties. The concept of morality I wish to defend is this:

> x has a morality or moral AG only if it includes judgements, rules, principles, ideals, etc., which (d) concern the relations of one individual (e.g. x) to others, (e) involve or call for a consideration of the effects of his actions on others (not necessarily all others), not from the point of view of his own interests or aesthetic enjoyments, but from their own point of view.[2]

Two or three notes are needed here. (1) I do not mean to say that an AG is a morality if it is social in this sense. I am suggesting that being social in this sense is a necessary, not a sufficient condition, of being a morality. In fact, I think that Hare's formal criteria of prescriptivity and universalizability are also necessary, and perhaps others – perhaps also supremacy, but of that more later. (2) My proposed condition is a relatively non-committal one. It admits as moralities deontological AGs as well as utilitarian ones, nationalistic and class ideologies as well as universalistic

[1] See op. cit., pp. 168 f., for his mention of the third. [2] Op. cit., p. 66.

ones, justice as well as benevolence, inequalitarian as well as equalitarian theories of justice, etc. It rules out as non-moral only such AGs as pure egoism or prudentialism, pure aestheticism, and pure religion. But this checks with much actual usage. We often contrast a prudential or aesthetic point of view with a moral one, and many theologians refer to a purely religious AG as 'transmoral' or 'beyond morality', as Tillich does. My criterion does not even rule out the Nazi AG, which Hare insists must be considered as a morality, for, though the Nazi AG did not call for a consideration of Jews, it did require an individual to consider other Germans. It must be remembered in this connection that, in saying the Nazi AG is a morality, Hare and I are not saying that it is a valid moral code or that it is right to act on it. 'The Case of Nietzsche' must also be mentioned again, for it is often cited as an argument against social definitions of morality. Has Nietzsche a morality? My answer is that this depends on how one reads him. If one takes his proposed AG as a purely egoistic or aesthetic one, then I say it is not a morality, but then he himself sometimes thinks of it as 'beyond good and evil' or as anti-moral. But one can also interpret Nietzsche as calling for one 'master' to respect others as equals and even to help them on their way. Thus spake Zarathustra: 'Be unto thy neighbour an arrow, and a longing for the superman.' Taken in this way, Nietzsche has a right to call his AG 'master-morality' even on my view and need not be read out of the moral party. (3) It should be remarked here that it is only in a sense that any principle, ideal, or way of life is 'built out of' morality by any definition. There is a sense in which even a utilitarian definition of morality does not exclude egoism or aestheticism. For, even if one takes a concern for the general good to be part of the definition of morality, one may still hold that doing what is for one's own good is for the greatest general good, and that, therefore, one should take as one's working criterion of morality the promotion of one's own good. Whether egoism as a working criterion is excluded by a social definition or not depends on the facts about the world. A social definition of morality is not as such incompatible with any substantive working criterion; it rules out only views that build egoism, aestheticism, etc., into the definition of morality (and to this there can be no objection) and views that take egoism, aestheticism, etc., as the basic principles or ideals of their AGs.

As indicated above, my presentation will consist mainly of a discussion of points made by Hare and Falk, but perhaps I should try to say briefly why I am troubled by non-social definitions of the kind they favor. (1) They regard a social definition as 'misleading'. To me, however, it seems at least as misleading to say than an AG is a moral one provided only that it satisfies the wider formal conditions formulated earlier. If saying that an AG is not a morality suggests that it is wicked or at least negligible, saying that it is a morality suggests that it is justified and ought to be supported by the moral sanctions of society. If saying that an 'ought' is non-moral suggests that it is questionable or unimportant, saying that it is moral suggests that it is socially important and legitimate. I am somehow more troubled by the latter misleadingness than by the former. (2) Falk insists that we must not 'expect morality on all levels to do the same kind of job as the institution of law', and I agree that morality is and should be different from law – in such ways as Hart has spelled out.[1] But I still share Kemp's feeling that, 'if by the "function" of a practice is meant the reason why it exists and is carried on' (Falk's words), then the function of morality is not just to serve as a supreme action-guide but to make possible 'some kind of cooperation or social activity between human beings'.[2] (3) Opponents of a social definition object that it refuses to call some seriously adopted ways of life moralities. I, however, am reluctant to see them all thought of as forms of morality – aestheticism, prudentialism, or even 'immoralism' (whatever that is). (4) The wider, formal definition of morality, by casting its net so widely, seems to me to blur and relativize any distinction between morality and other things like prudence, religion, etc.: any of these other AGs may become moralities by being taken seriously enough, being universalized, etc. This makes it very difficult to discuss questions about the relations of morality to other things. (5) On the wider conception of morality, a man's moral reasoning is complete if he brings his proposed action under an AG which he regards as

[1] Cf. H. L. A. Hart, *The Concept of Law*, 1961, pp. 169–76.

[2] See Falk, op. cit., p. 63; J. Kemp, op. cit., p. 196.

definitive for his conduct (and perhaps also for that of others), even if he does not consider others in any important way. This strikes me as in itself paradoxical, but it also makes it unclear just what moral justification society could have for punishing or restraining any individual who is acting 'conscientiously' in accordance with his own AG, whatever this may be and whatever his action is.

IV

Coming now to Hare's discussion of our question, we must remember that – as compared with Aiken, MacIntyre,[1] and others – he is ready to build *something* into the definition of 'moral' and 'morality', namely, prescriptivity and universalization (and perhaps supremacy). Thus, if A says in opposition to B, 'I ought to put my debtors into prison' but is either 'not using it prescriptively, or not universalizably', then Hare is willing to say (1) that A's judgement is not a moral one, (2) that A and B are not in 'substantial moral disagreement', or that 'the apparent moral disagreement is really only verbal', and (3) that there may be a residuum of substantial disagreement between them but this 'cannot be moral' or even evaluative. However, Hare is averse to building 'any particular moral opinion' into the meaning of 'moral' or 'morality', which is what he regards social and utilitarian definitions as doing, and hence, he refuses to condone a similar move against someone who says, 'I ought to put my debtors into prison', but denies that social considerations are relevant to his judgement. He opposes limiting the terms 'moral judgement' and 'moral question' to 'questions concerning the effects of our actions upon other people's interests' by a 'terminological fiat'. This may be partly because he thinks that a terminological fiat in this case is 'arbitrary' in a way in which his own is not, but in view of the amount of support there is for a social definition in ordinary use, this is hardly plausible – and can hardly be decisive anyway. Besides, as I have sought to suggest, there are reasons for making such a fiat. More basic, for Hare, no doubt, is

[1] A. C. MacIntyre, 'What Morality Is Not', *Philosophy*, 1958 (reprinted in this volume). *Eds.*

his wish to be as non-committal as possible in the definition of morality, with which I have already indicated some sympathy.[1]

In his first discussion of the latter move,[2] Hare himself points to a strong reason for making it, namely, that it makes possible a 'neat solution' of the problem of delimiting moral from other evaluative judgements and questions (which is just what I would like). But he rejects the move anyway on two grounds: (a) that 'ordinary people commonly apply the name "moral" ' to a class of questions that do not involve the interests of other people,[3] (b) 'that these other questions [should] be dealt with inside moral philosophy, and not neglected'. He gives three examples of such questions: (1) whether it is wrong for a pretty girl to earn money by undressing herself at a 'strip club' for the pleasure of an audience of middle-aged business men, (2) whether one should make one's career as a stockbroker or as an army officer, when it makes no appreciable difference to other people's interests which he chooses, and (3) whether the leader of a Himalayan expedition should lead the final assault on the mountain himself or give someone else the opportunity, it being supposed that the interests of those involved are equally served in either case.

In comment on these cases, I want to agree that the first question, as stated, is a moral one, since the strip teaser's conduct would have effects on other people's lives, directly or indirectly. But suppose it would not have such effects. Would ordinary people, knowing this, still call the question moral? This I doubt, and Hare does nothing to show that they would. In fact, he himself indicates that they would then use such quasi-aesthetic words as 'degrading', and this strikes me as a good reason for distinguishing such a judgement from one that is strictly moral.

Hare takes all three questions, not as questions involving people's interests, but as questions involving only purely personal ideals of what one should be. I have no wish to denigrate the importance of such ideals, any more than Hare and Falk do. Nor do I wish to exclude ideals from morality. Hare implies that, if such purely personal ideals are regarded as non-moral, then moral philosophers must ignore all questions about ideals. But

[1] For phrases quoted see Hare, pp. 98, 192, 146 f. [2] Pp. 146–52.
[3] Cf. Hart, op. cit., pp. 164 f., 170, 177.

this is to forget that there may also be ideals about one's conduct in relation to others, e.g. the ideal of the just man, which are, of course, moral, and certainly should be discussed by moral philosophers. The question is about the purely personal ideals, and I I do not see that Hare shows that we either do or should call them 'moral'. He suggests that we should be at a loss for a word to describe them if we did not call them 'moral', but I see no difficulty in thinking of them as personal rather than moral. When one says, 'I know my own worth' or 'I put a high value on myself', one is not, or at least not necessarily, making a moral judgement. Indeed, Hare himself remarks that questions involving purely personal ideals are 'very like aesthetic ones – as if a man were regarding his own life and character as a work of art'. I agree. He also admits that, if we call them 'moral', it may be that this use of the word is not the same use as when we say that the question whether one should put creditors in prison is a moral one. But if all this is true, it seems to me that there is good reason for not regarding them as properly moral questions, even if ordinary people do so regard them.

I should mention here that Falk seems to me to do much better than Hare by way of arguing that purely 'personal' oughts and ideals should be called 'moral'. He does show that they are 'at least akin' to moral oughts and ideals 'in their action-guiding force and function'. But even he adds, '. . . I grant that one does not want to speak of more than a kinship', and so stops short of actually taking them as fully moral.[1]

Hare is concerned to maintain that moral philosophers should say something about purely personal ideals, and he seems to believe that if morality is defined as I suggest it should be, then moral philosophers, *qua* moral philosophers, are debarred from discussing such ideals. I certainly want to insist that philosophers should discuss personal ideals under 'the philosophy of life' – something they too seldom do. I doubt, however, that they should deal with them under the heading of strictly moral questions, as if all ideals were moral. But, in any case, there is at least one ground on which moral philosophers as such may quite appropriately discuss personal ideals. They may and should ask whether

[1] Op. cit., p. 38.

the pursuit of a given personal ideal is compatible with morality, whether personal ideals ever take precedence over moral principles in cases of conflict, and so on.

In his next treatment of our question,[1] Hare asks whether we may distinguish moral judgements from aesthetic and other value-judgements (and give the 'liberal' the victory over the 'Nazi') by building some kind of consideration of other people into the definition of 'moral judgement'. Then he writes:

> To put this restriction upon the use of the word 'moral' is to write some kind of utilitarianism into its definition. It is fortunately not necessary to inquire precisely how the definition would be formulated, and whether this could be done without making it naturalistic . . . for not only is any such attempt open to a fatal objection, but, as we shall see, we can achieve all that we need to achieve by less questionable means.

Hare's first point here seems to be that a social definition of morality builds utilitarianism into the definition of morality, thus begging the question whether the principle of utility is the criterion of morality or not. I have already tried to mitigate the force of this objection by proposing a social definition which does not rule out deontological views of the moral standard and so does much less question-begging than Hare thinks. I have even pointed out that there is a sense in which it does not rule out any working criterion of morality. It precludes only views to the effect that egoism (e.g.) is not just the working criterion of morality but its basic principle. And why should such views not be written out by definition? Because saying that egoism is the *basic* principle of morality and not merely its working criterion is a moral judgement and hence must not be defined as non-moral? But *is* saying this a moral judgement? It looks more like a logical one to me. Hare also implies that a social definition of morality involves a kind of 'naturalism'. This is interesting and has been suggested before – e.g. by MacIntyre, who even accuses Hare of naturalism because Hare defines morality in terms of impersonality or universalizability. Well, what is so bad about naturalism? 'The most fundamental objection against naturalism',

[1] Pp. 162–5.

says Hare, 'is that it makes moral questions depend upon conceptual ones';[1] it writes a 'particular moral opinion' into the definition of 'right', 'good', 'moral', etc., and (in effect) seeks to settle the issue between that opinion and opposing ones by conceptual or terminological legislation. Hare is thinking that if one defines 'moral judgement' so as to include a consideration of the interests of others, then one is committing oneself to regarding a neglect of other people's interests as immoral (morally wrong) and not merely as non-moral. He does not, however, regard his own definition of morality as similarly committing him to any moral opinion whatsoever, e.g. that the Nazi's treatment of Jews is wrong, because he argues that no substantive moral opinion is ruled out by his formal criteria alone. But if what was said earlier is correct, then even my more material definition of morality does not as such commit me to saying that pursuing one's own best interests is morally wrong, that acting on aesthetic considerations is morally wrong, etc., since it might still be that pursuing one's own best interests, etc., is for the welfare of everyone. Does it commit me to any particular moral judgement?

In reply, notice first that my saying that an AG is a morality only if it requires one to guide one's action at least in part by a certain kind of consideration for the lives of others does not commit me to saying that not doing so at all is wrong. At most it commits me to saying that not doing so at all is morally wrong, wrong on moral grounds, or from a moral point of view. And, indeed, it does look as if my social definition of morality does commit me to saying this or something like it.

Notice too that my saying that an AG is a moral one does not commit me to saying that acting on it is morally right. For I may hold that it is a moral AG without myself accepting it as the standard of moral right and wrong. In saying that it is a moral AG only if it requires its subscribers to consider others in certain ways, however, I am asserting that one cannot claim one's judgements to be moral unless they include such a consideration of others, and this seems to mean that one cannot judge anything to be morally right or morally wrong unless one considers others

[1] P. 187.

in the required ways, and that it is morally wrong not to consider them in those ways at all, i.e. living by a non-moral AG is immoral.

What is objectionable in such 'naturalism'? Hare's answer is that it 'makes moral questions depend upon conceptual ones'. Suppose that you say, 'One ought to disregard other people's interests completely' and I counter with 'No, that would be wrong', defending my reply by citing my social definition of morality and proclaiming that your thesis just is not a moral one. This move is what Hare objects to, and I agree that I cannot finesse you in this manner. (a) You might answer, 'What has that got to do with it? I was not meaning to make a moral judgement. But I still maintain that one ought to disregard others.' In this case, you accept my definition of 'moral' and agree that your dictum is non-moral, but you insist on its substantial correctness. And, of course, I cannot settle the substantial issue between us (about how we should live) by any conceptual or terminological legislation, certainly not by a modest proposal for using the term 'moral' such as I am advancing. But it still remains true that *if* you accept my definition, you cannot claim that it is morally right to disregard our neighbors entirely. (b) You may reject my definition and insist that you were making a moral judgement. Then, following a long custom, you might use the 'open question' argument. You might argue, e.g. that it is not a contradiction to say, 'A total disregard for others is right from a moral point of view' or that it is not silly to ask, 'Is a total disregard for others wrong from a moral point of view?' However, this famous argument does not seem to me to be at all conclusive. In the first place, while it is clear to me that 'This involves an utter disregard for other people's interests, but is it wrong?' is an open question, it is not clear to me that 'This involves an utter disregard for other people, but is it wrong on moral grounds?' is also an open question when one remembers that the issue is not the meaning of 'wrong' but that of 'moral'. Second, even if it is an open question, this would at most show that there is a wider use of 'moral' besides the social one; it would do nothing to refute a proposal to limit the word 'moral' to its narrower, social use.

It is here that Hare's 'fatal objection' comes in.

> The objection is this. If we give the liberal the victory over the Nazi by means of a definition of this kind, the victory will be in an important sense barren. Both will be left prescribing, universally, different ways of life, and therefore differing about the most fundamental questions that people can differ about. It is not of fundamental importance whether we *call* the difference between them a moral difference, as most people would. . . . This is a purely terminological question, to which we shall return. What does matter is that the difference between these people should be recognized to exist and that we should do all we can, as moral philosophers, to promote clear thinking about it.

Frankly, however, I see no fatal objection in this passage as it stands; in fact, there is nothing in it with which one cannot agree heartily. I agree that adopting a social definition of morality will not settle the substantial issue between, say, the Nazi and the liberal, which is between two profoundly differing ways of life, and that what matters is to understand this issue and if possible, to resolve it. But our question here is whether we should adopt a social definition of the word 'moral' or not, and on this question the passage quoted only refers us to a later discussion, to which we shall come in due course.

It seems implicit in Hare's words, however, that he is thinking (a) that we do and should call the difference between the liberal and the Nazi a 'moral' one, perhaps because it is so fundamental, (b) that we cannot do so if we adopt a social definition of 'moral', and (c) that the issue ought to be discussed by moral philosophers as such. I agree in wanting to call the issue between the liberal and the Nazi (or between Toulmin and Nietzsche, etc.) a moral one, and in believing that moral philosophers ought to discuss it. But is this incompatible with a social definition of morality? Hare assumes that the issue between the liberal and the Nazi can be a moral issue only if both parties are making moral judgements or advocating moral principles. But suppose that X is advocating the moral way of life, whatever that is, and Y is advocating some other way of life. Then by hypothesis Y is not advancing a moral principle, but it still seems to me quite appropriate to call the issue a moral one, since it is about morality. On a social definition of morality, then, even if the liberal's AG is a moral one and the Nazi's is not, one can still call the issue 'moral' because it amounts to a debate about whether we should be moral or not. Certainly this is an issue that moral philosophers as such should discuss – as, with the possible exceptions of Prichard and Toulmin, they always have (including Baier, who accepts a social definition of morality).

Hare goes on, however. He again raises the question whether we should build a consideration of other people's interests into the definition of 'moral' (Does this mean that he has not yet stated his fatal objection?). Then he writes,

> Is this merely a question of how the word 'moral' is actually used in common speech: should we or should we not *call* the Nazi's position a moral one? It cannot be emphasized too strongly that the question that is troubling us is not of this sort. . . . Nothing in mathematics turns on the actual use of words, and neither does it in morals. We could get along just as well if we used artificial symbols for all the concepts involved. . . . Let us for the moment abandon the word 'moral' as too slippery, and set out the concepts involved more fully.[1]

Hare then distinguishes three concepts: $ought_1$, which is bound by a requirement to consider other peoples' interests, $ought_2$, which is not bound by this restriction, but only by those of prescriptivity and universalizability, and $ought_3$, which is prescriptive but not universalizable. In these terms the difference between Hare and myself may be expressed thus: I propose to identify the moral ought with $ought_1$, while he (apparently) proposes to identify it with $ought_2$. He does not scruple about calling $ought_3$ 'non-moral' any more than I do (except that he doubts 'ought' is ever used in that sense). Neither of us is concerned simply about the actual use of 'moral'. But Hare does not proceed to debate this issue; he forgets about $ought_1$ and tries to clarify the dispute between the liberal and the Nazi without using it, i.e. by using only $ought_2$.

We must, therefore, look at Hare's promised later discussion of the use of 'moral'.[2] His main points there seem to be: (1) to identify the moral ought with $ought_2$ does not commit one to any particular moral opinion, (2) to identify it with $ought_1$, however,

[1] P. 164. [2] Pp. 191–202.

does commit one to a certain kind of moral doctrine, and (3) it leaves one with no way of expressing the issue between oneself and one's opponents. These points have already been discussed, but I must add one remark about the third. Hare appears to think that, if one defines the moral ought as $ought_1$, one not only cannot refer to the issue between oneself and one's opponents as a moral one, a point already dealt with, but one also cannot even describe the issue as a normative or evaluative one. This, however, is a mistake, for one can still allow that there are other, non-moral oughts, and hence, that the question, 'Why be moral?' makes sense and is a normative issue.

Earlier we quoted Hare as claiming that 'we can achieve all that we need to achieve by less questionable means' than by building any kind of sociality into the definition of 'moral'. And he does, indeed, show that we can state and discuss all substantive issues without resorting to a social definition of morality. However, he does not show us how we can distinguish moral judgements from aesthetic and other kinds of value-judgements. He seems to forget that this is part of what we wanted to achieve. He writes in one place that, 'The enormity of Nazism is that it extends an aesthetic style of evaluation into a field where the bulk of mankind think that such evaluations should be subordinated to the interests of other people.'[1] But his own refusal to define morality by reference to the interests of other people permits just such an extension of an aesthetic style of evaluation to take place. That is one of the reasons why I am unhappy with his wide conception of the field of morality.

We may conclude, then, that we need not give Hare the victory in his running battle with the proponents of a narrower, social concept of morality.

V

Falk, in his penetrating paper, makes what seems to me a much better case for adopting the wider, non-social, 'formalist' conception of morality. It would be most rewarding, if there were time, to review his discussion in detail. Still, though he seems to

[1] P. 161.

favour the wider concept, he does not actually offer us a 'ruling' on how we are to use the term 'moral'. In fact, at the crucial points he appears almost to accept the narrower, social concept – with two provisos (a) that we admit 'that there are other than strictly moral commitments which a right-living person may have to reckon with no less than his strictly moral ones', and (b) that we face the implications of our definition.[1] If this is his actual conclusion, then I may even regard him as an ally, albeit a reluctant one. (It may be, however, that his strategy is to bring us 'non-formalists' to the point of admitting this proviso and facing these implications, hoping that we will then prefer to shift to the wider definition of a morality as any reasoned body of action-guiding commitments that is taken as an authoritative and supreme rule of life.) Hence, I shall not review his discussion as I have Hare's, but confine myself to a reaction to what he says in his last section,[2] which seems to me profoundly interesting in connection with our problem.

In this section Falk distinguishes and contrasts two criteria of morality. One is essentially that of Jaspers – that a moral AG is one which is taken as an authoritative and supreme rule of life. The other is a social criterion of some kind such as I have been supporting. These two criteria, he says, are both involved in our ordinary discourse about morality, and one can define primary or group morality by using both of them – as 'authoritative action-guidance whose function is to regulate the social order'.

> But morality, on the mature level, is less well-conceived in this way. There are difficulties in uniting the authoritative and the social associations of 'moral' in one concept.

There are mature, reflective, autonomous AGs and the interesting question is, 'When is such an AG a morality?'

> There are those who insist that mature morality is socially beneficial ought-abidance . . . [prescribing] a material as well as a formal criterion for the use of 'moral'. There are others who call 'moral' any definitive and 'authentic' commitment of a self-directing person, whether its grounds are social or personal.

[1] Cf. Falk, op. cit., pp. 30, 48, 53, 65. [2] Ibid., pp. 60–6.

If we take the second alternative, we must accept the implication that a morality will not necessarily be socially beneficial or considerate; if we take the first, we must accept the implication that a morality may not be definitive or supreme in life, for a man may not find the case for being social rationally convincing.

> The moral and the definitive commitments on the mature level need not then coincide . . . 'morality' on this level is demoted from its accustomed place of being the sole and final arbiter of right and wrong choice.

Here, Falk says, is an issue in which there is a semantic case for each side, but which it is 'far more important to understand than to take sides on'. And then comes his last paragraph.

> If both alternatives are repugnant, it is because both fall short of expectations. The unequivocal successor to the primary moral law should be a commitment by noncoercive reasons, manifestly binding on everyone alike, to give precedence always to the claims of beneficence and the requirements of social living. But there is no warrant for assuming such a commitment on the level of autonomous choice. The rules of language cannot furnish it any more than pure reason, or intuition. The hard fact is that the rational and autonomous mode of life overlaps, but no longer necessarily coincides, with the moral mode of life as conceived from the point of view of the social interest. The autonomous agent can be a debatable social asset. It is vain to expect morality on all levels to do the same kind of job as the institution of the law. The concept of morality itself bears the accumulated scars of conceptual evolution. Its multiple associations are a bar to summing it up in any one way.[1]

This is a puzzling, almost despairing, paragraph. Falk seems to be saying that we can about equally reasonably take either side, provided that we know what we are doing, but he finds both alternatives 'repugnant', and sees no way of combining them; and so, regarding discretion as the better part of valour, he concludes that 'clarification is a safer bet than decision'.

However, I am foolhardy enough to want some kind of decision. And so far, I have been defending the inclusion of a social criterion in any definition of morality. But, as a matter of fact, I

[1] Ibid., p. 33.

have always also been inclined to think of morality as the final and supreme tribunal in matters of conduct. Hence, I shall conclude by proposing that we continue to include both 'the authoritative and the social associations of "moral"' in our concept of morality, as we have in the past, i.e. that we define a moral AG roughly (using Falk's words) as 'a commitment by noncoercive reasons, manifestly binding on everyone alike, to give precedence always to the claims of beneficence and the requirements of social living'.

There is one consideration, suggested by Falk himself,[1] which makes me hesitate to include the notion of definitiveness or supremacy in the definition of a morality. This is the fact that we sometimes seem rightly to give non-moral considerations priority over moral ones. Thus, we may say to a daughter tied to a demanding mother, 'You owe it to yourself to break away, hard as it may be on your mother.' Or we may say of an artist, 'His moral peccadillos are more than compensated for by his achievement of self-expression.'[2] It may be, however, that when we speak thus we are still vaguely thinking that morality itself permits such exceptions to its dictates and is not renouncing its claim as ultimate arbiter. Certainly we are still thinking that there remain points at which its claims are supreme. Perhaps, then, we need not take such cases as showing that we should not build over-ridingness into our definition of a moral AG.

But, of course, there is still the difficulty pointed out by Falk, namely, the fact that a socially considerate AG is not necessarily one which a mature person will find rationally authoritative. It is this fact that leads some theologians to advocate our going 'beyond morality' into 'religion'. Short of doing this, the only alternative is to postulate a 'coincidence between ought-abiding living and the social interest' or between the requirements of final rationality and those of social considerateness.

Morality as we have known it in our history, with its dual emphasis on rationality and sociality, represents, as it were, a bet that man and the world are such that these two requirements do coincide, and my proposal really is that we build this bet into our future concept of morality too. The coincidence in question

[1] Ibid., p. 49. [2] I owe this example to J. O. Urmson.

is not something that can be proved in our present state of knowledge, but neither is it something that has been disproved. It may and must be postulated. Kant thought that this postulate must take the form of a belief that there is a God who sees to it that a life of moral virtue is rewarded by the happiness it deserves. But it need not take this form. In some form or other it is made by Plato and Aristotle, by Butler and Sidgwick, and by all of the religious and idealistic thinkers who believe that he who loses his life for morality's sake shall gain it. Even John Dewey postulates a coincidence of a kind between the happy life and the socially considerate and interested one, and, though Reinhold Niebuhr labels Dewey's faith naive and touching, he himself similarly believes that love is the true way to self-realization. In fact, I cannot help but feel that Falk and Hare, even when they are most in favor of the formalist view, are in a sense betting that the AG which a fully rational and informed man would choose under these criteria is also one which would be socially desirable. Would they actually regard the wider concept of morality as acceptable if they did not believe this? For example, would Hare be content with his formal criteria if he did not believe that, in applying them, he can show that man and the world are such that only a 'fanatic' can be a Nazi?

I have no wish to sound mystical. It does not seem to me that the postulate is at all plausible if it says that an individual is never a loser in prudential terms – i.e. in terms of his happiness in this life – by doing the socially moral thing. But is the finally rational way of life for the individual to be identified with that which is for his own greatest happiness? Butler and Kant thought that it was, but Falk seems to me to be on a better track when he insists that rational personal oughts need not take a prudential form.[1] The finally rational course, he suggests, is 'what a man would do in his wisdom – if he were to consider things widely, looking past the immediate concerns of self and giving essentials due weight before incidentals', and it must not simply be assumed that this is the course known as egoism. Even if we identify being rational with calm deliberation in the light of full knowledge about what one wants and how to get it, it does not follow that the rational

[1] Falk, op. cit., pp. 33–8.

life is that of cool self-love. Whether it is or not depends on what one wants when one is enlightened about himself and his world. One might then want exactly the kind of world that a socially considerate AG would call for, even at the cost of some sacrifice on one's own part. There is some evidence in modern psychology – so Erich Fromm and others have been arguing – that this is in fact the case. At least it may be that as our insight into man, society, and the universe increases we shall more and more come to see that the finally rational way of life for the individual is or at least may be precisely the socially considerate one. It is true, as Niebuhr and others have been dinning into our ears, that there is a good deal of evidence of something 'demonic' in human beings, something anti-social, selfish, proud, sensual, wild, or what not. This may make it hard to be optimistic about the future of morality, but it does not mean that we cannot postulate the congruence of the rational and the moral life. For such a demonic element, if it exists, is just as anti-rational as it is anti-moral. It may be, of course, that the congruence holds only for those who are in some sense 'reborn' – in Dewey's words, only for those who have come to choose 'objects that contribute to the enrichment of the lives of all'[1] – and then we must hope that such a rebirth is possible, and not for a few only.

VI

That is rather vague talk, but it is the best I have to offer at the moment. It is intended to do a little to make acceptable a proposal to define a moral code as one that calls for at least a certain minimum of consideration of others and at the same time is taken as rationally supremely authoritative. In this proposal, the narrower social and wider formal concepts of morality are combined, with the help of a postulate. The other chief alternatives are: (1) to give up altogether all talk of 'morality' and the 'moral', (2) to think in terms of two (or more) concepts of morality, one formal and wider, the other social and narrower, (3) to keep only the wider notion, and (4) to keep only the narrower one. If the postulate referred to

[1] Cf. J. Dewey, *Theory of the Moral Life*, 1960, pp. 167 f., 104

is dubious or false, and I am forced to choose between these other alternatives, then I should choose the last. The first two seem rather obviously unsatisfactory, and I have tried to indicate why I am unhappy with the third. It is true that the last alternative leaves us with the question whether it is rational to adopt a moral AG, but I am willing to wrestle with this question if I must. Still, I hope I do not have to, for I would like to believe that morality is a rational enterprise as well as a social one.

10

Moral arguments

PHILIPPA FOOT

Those who are influenced by the emotivist theory of ethics, and yet wish to defend what Hare has called 'the rationality of moral discourse', generally talk a lot about 'giving reasons' for saying that one thing is right, and another wrong. The fact that moral judgements need defence seems to distinguish the impact of one man's moral views upon others from mere persuasion or coercion, and the judgements themselves from mere expressions of likes and dislikes. Yet the version of argument in morals currently accepted seems to say that, while reasons must be given, no one need accept them unless he happens to hold particular moral views. It follows that disputes about what is right and wrong can be resolved only if certain contingent conditions are fulfilled; if they are not fulfilled, the argument breaks down, and the disputants are left face to face in an opposition which is merely an expression of attitude and will. Much energy is expended in trying to show that no sceptical conclusion can be drawn. It is suggested, for instance, that anyone who has considered all the facts which could bear on his moral position has *ipso facto* produced a 'well founded' moral judgement; in spite of the fact that anyone else who has considered the same facts may well come to the opposite conclusion. How 'x is good' can be a well founded moral judgement when 'x is bad' can be equally well founded it is not easy to see.

Reprinted from Mind, *1958, pp. 502–13, by permission of the author and* Mind. *This paper has been discussed in A. Stigen, 'Mrs Foot on Moral Arguments',* Mind, *1960, and R. Robinson, 'Argument and Moral Argument',* Mind, *1961.*

The statement that moral arguments 'may always break down' is often thought of as something that has to be accepted, and it is thought that those who deny it fail to take account of what was proved once for all by Hume, and elaborated by Stevenson, by Ayer, and by Hare. This article is an attempt to expose the assumptions which give the 'breakdown' theory so tenacious a hold, and to suggest an alternative view.

Looked at in one way, the assertion that moral arguments 'may always break down' appears to make a large claim. What is meant is that they may break down in a way in which other arguments may not. We are therefore working on a model on which such factors as shortage of time or temper are not shown; the suggestion is not that A's argument with B may break down because B refuses for one reason or another to go on with it, but that their positions as such are irreconcilable. Now the question is; how can we assert that any disagreement about what is right and wrong may end like this? How do we know, without consulting the details of each argument, that there is always an impregnable position both for the man who says that X is right, or good, or what he ought to do, and for the man who denies it? How do we know that each is able to deal with every argument the other may bring?

Thus, when Hare describes someone who listens to all his adversary has to say and then at the end simply rejects his conclusion, we want to ask 'How can he?' Hare clearly supposes that he can, for he says that at this point the objector can only be asked to make up his mind for himself.[1] No one would ever paint such a picture of other kinds of argument – suggesting, for instance, that a man might listen to all that could be said about the shape of the earth, and then ask why he should believe that it was round. We should want, in such a case, to know how he met the case put to him; and it is remarkable that in ethics this question is thought not to be in place.

If a man making a moral judgement is to be invulnerable to criticism, he must be free from reproach on two scores: (*a*) he must have brought forward evidence, where evidence is needed; and (*b*) he must have disposed of any contrary evidence offered.

[1] *The Language of Morals*, p. 69.

It is worth showing why writers who insist that moral arguments may always break down assume, for both sides in a moral dispute, invulnerability on both counts. The critical assumption appears in different forms because different descriptions of moral arguments are given; and I shall consider briefly what has been said by Stevenson and by Hare.

I. Stevenson sees the process of giving reasons for ethical conclusions as a special process of non-deductive inference, in which statements expressing beliefs (R) form the premises and emotive (evaluative) utterances (E) the conclusion. There are no rules validating particular inferences, but only causal connections between the beliefs and attitudes concerned. 'Suppose,' he writes, 'that a theorist should *tabulate* the "valid" inferences from Rs to Es. It is difficult to see how he could be doing anything more than specify what Rs he thereby resolves to *accept* as supporting the various Es. . . . Under the name of "validity" he will be selecting those inferences to which he is psychologically disposed to give assent and perhaps inducing others to give a similar assent to them.'[1] It follows that disputes in which each man backs up his moral judgement with 'reasons' may always break down, and this is an implication on which Stevenson insists. So long as he does not contradict himself and gets his facts right, a man may argue as he chooses, or as he finds himself psychologically disposed. He alone says which facts are relevant to ethical conclusions, so that he is invulnerable on counts (*a*) and (*b*): he can simply assert that what he brings forward is evidence, and can simply deny the relevance of any other. His argument may be ineffective, but it cannot be said to be wrong. Stevenson speaks of ethical 'inference' and of giving 'reasons', but the process which he describes is rather that of trying to produce a result, an attitude, by means of a special kind of adjustment, an alteration in belief. All that is needed for a breakdown is for different attitudes in different people to be causally connected to the same beliefs. Then even complete agreement in belief will not settle a moral dispute.

II. Hare gives a picture of moral reasoning which escapes the difficulties of a special form of inference without rules of validity.

[1] *Ethics and Language*, pp. 170–1.

He regards an argument to a moral conclusion as a syllogistic inference, with the ordinary rules. The facts, such as 'this is stealing', which are to back up a moral judgement are to be stated in a 'descriptive' minor premise, and their relevance is to be guaranteed by an 'evaluative' major premise in which that kind of thing is said to be good or bad. There is thus no difficulty about the validity of the argument; but one does arise about the status of the major premise. We are supposed to say that a particular action is bad because it is a case of stealing, and because stealing is wrong; but if we ask why stealing is wrong, we can only be presented with another argument of the same form, with another exposed moral principle as its major premise. In the end everyone is forced back to some moral principle which he simply asserts – and which someone else may simply deny. It can therefore be no reproach to anyone that he gives no reasons for a statement of moral principle, since any moral argument must contain some undefended premise of this kind. Nor can he be accused of failing to meet arguments put forward by opponents arguing from different principles; for by denying their ultimate major premises he can successfully deny the relevance of anything they say.

Both these accounts of moral argument are governed by the thought that there is no logical connection between statements of fact and statements of value, so that each man makes his own decision as to the facts about an action which are relevant to its evaluation. To oppose this view we should need to show that, on the contrary, it is laid down that some things do, and some things do not, count in favour of a moral conclusion, and that a man can no more decide for himself what is evidence for rightness and wrongness than he can decide what is evidence for monetary inflation or a tumour on the brain. If such objective relations between facts and values existed, they could be of two kinds; descriptive, or factual premises might *entail* evaluative conclusions, or they might count as *evidence* for them. It is the second possibility which chiefly concerns me, but I shall nevertheless consider the arguments which are supposed to show that the stronger relationship cannot exist. For I want to show that the arguments usually brought forward do not *even* prove this. I want to say that it has not even been proved that moral conclusions cannot be entailed by factual or descriptive premises.

It is often thought that Hume showed the impossibility of deducing 'ought' from 'is', but the form in which this view is now defended is, of course, that in which it was rediscovered by G. E. Moore at the beginning of the present century, and developed by such other critics of 'naturalistic' ethics as Stevenson, Ayer and Hare. We need therefore to look into the case against naturalism to see exactly what was proved.

Moore tried to show that goodness was a non-natural property, and thus not to be defined in terms of natural properties; the problem was to explain the concept of a 'natural property', and to prove that no ethical definition in terms of natural properties could be correct. As Frankena[1] and Prior[2] pointed out, the argument against naturalism was always in danger of degenerating into a truism. A natural property tended to become one not identical with goodness, and the naturalistic fallacy that of identifying goodness with 'some other thing'.

What was needed to give the attack on naturalism new life was the identification of some deficiency common to the whole range of definitions rejected by Moore, a reason why they all failed. This was provided by the theory that value terms in general, and moral terms in particular, were used for a special function – variously identified as expressing feelings, expressing and inducing attitudes, or commending. Now it was said that words with emotive or commendatory force, such as 'good', were not to be defined by the use of words whose meaning was merely 'descriptive'. This discovery tended to appear greater than it was, because it looked as if the two categories of fact and value had been identified separately and found never to coincide, whereas actually the factual or descriptive was defined by exclusion from the realm of value. In the ordinary sense of 'descriptive' the word 'good' is a descriptive word and in the ordinary sense of 'fact' we say that it is a fact about so and so that he is a good man, so that the words must be used in a special sense in moral philosophy. But a special philosopher's sense of these words has never, so far as I know, been explained except by contrasting

[1] W. K. Frankena, 'The Naturalistic Fallacy', *Mind*, 1939.

[2] A. N. Prior, *Logic and the Basis of Ethics*, chap. I

value and fact. A word or sentence seems to be called 'descriptive' on account of the fact that it is *not* emotive, does *not* commend, does *not* entail an imperative, and so on according to the theory involved. This might seem to reduce the case against naturalism once more to an uninteresting tautology, but it does not do so. For if the non-naturalist has discovered a special feature found in all value judgements, he can no longer be accused of saying merely that nothing is a definition of 'good' unless it is a definition of 'good' and not 'some other thing'. His part is now to insist that any definition which fails to allow for the special feature of value judgements must be rejected, and to label as 'naturalistic' all the definitions which fail to pass this test.

I shall suppose, for the sake of argument, that the non-naturalist really has identified some characteristic (let us call it f) essential to evaluative words; that he is right in saying that evaluations involve emotions, attitudes, the acceptance of imperatives, or something of the kind. He is therefore justified in insisting that no word or statement which does not have the property f can be taken as equivalent to any evaluation, and that no account of the use of an evaluative term can leave out f and yet be complete. What, if anything, follows about the relation between premises and conclusion in an argument designed to support an evaluation?

It is often said that what follows is that evaluative conclusion cannot be deduced from descriptive premises, but how is this to be shown? Of course if a descriptive premise is redefined, as one which does not entail an evaluative conclusion, the non-naturalist will once more have bought security at the price of becoming a bore. He can once more improve his position by pointing to the characteristic f belonging to all evaluations, and asserting that no set of premises which do not entail an f proposition can entail an evaluation. If he takes this course he will be more like the man who says that a proposition which entails a proposition about a dog must be one which entails a proposition about an animal; he is telling us what to look out for in checking the entailment. What he is not so far telling us is that we can test for the entailment by looking to see whether the premise itself has the characteristic f. For all that has yet been shown it might be possible for a premise which is not f to entail a conclusion which is f, and it is obviously this proposition which the non-naturalist wants to deny.

Now it may seem obvious that a non-evaluative premise could not entail an evaluative conclusion, but it remains unclear how it is supposed to be proved.

In one form, the theory that an evaluative conclusion of a deductive argument needs evaluative premises is clearly unwarrantable; I mention it only to get it out of the way. We cannot possibly say that at least one of the premises must be evaluative if the conclusion is to be so; for there is nothing to tell us that whatever can truly be said of the conclusion of a deductive argument can truly be said of any one of the premises. It is not necessary that the evaluative element should 'come in whole', so to speak. If f has to belong to the premises it can only be necessary that it should belong to the premises *together*, and it may be no easy matter to see whether a set of propositions has the property f.

How in any case is it to be proved that if the conclusion is to have the characteristic f the premises taken together must also have it? Can it be said that unless this is so it will always be possible to assert the premises and yet deny the conclusion? I shall try to show that this at least is false, and in order to do so I shall consider the case of arguments designed to show that a certain piece of behaviour is or is not rude.

I think it will be agreed that in the wide sense in which philosophers speak of evaluation, 'rude' is an evaluative word. At any rate it has the kind of characteristics upon which non-naturalists fasten: it expresses disapproval, is meant to be used when action is to be discouraged, implies that other things being equal the behaviour to which it is applied will be avoided by the speaker, and so on. For the purpose of this argument I shall ignore the cases in which it is admitted that there are reasons why something should be done in spite of, or even because of, the fact that it is rude. Clearly there are occasions when a little rudeness is in place, but this does not alter the fact that 'rude' is a condemnatory word.

It is obvious that there is something else to be said about the word 'rude' besides the fact that it expresses fairly mild condemnation: it can only be used where certain descriptions apply. The right account of the situation in which it is correct to say that a piece of behaviour is rude is, I think, that this kind of behaviour causes offence by indicating lack of respect. Sometimes it is merely conventional that such behaviour does indicate lack of respect (e.g. when a man keeps his hat on in someone else's house); sometimes the behaviour is naturally disrespectful, as when one man pushes another out of the way. (It should be mentioned that rudeness and the absence of rudeness do not exhaust the subject of etiquette; some things are not rude, and yet are 'not done'. It is rude to wear flannels at a formal dinner party, but merely not done to wear a dinner jacket for tennis.)

Given that this reference to offence is to be included in any account of the concept of rudeness, we may ask what the relation is between the assertion that these conditions of offence are fulfilled – let us call it O – and the statement that a piece of behaviour is rude – let us call it R. Can someone who accepts the proposition O (that this kind of offence is caused) deny the proposition R (that the behaviour is rude)? I should have thought that this was just what he could not do, for if he says that it is not rude, we shall stare, and ask him what sort of behaviour would be rude; and what is he to say? Suppose that he were to answer 'a man is rude when he behaves conventionally', or 'a man is rude when he walks slowly up to a front door', and this not because he believes that such behaviour causes offence, but with the intention of leaving behind entirely the usual criteria of rudeness. It is evident that with the usual criteria of rudeness he leaves behind the concept itself; he may say the words 'I think this rude', but it will not on that account be right to describe him as 'thinking it rude'. If I *say* 'I am sitting on a pile of hay' and bring as evidence the fact that the object I am sitting on has four wooden legs and a hard wooden back, I shall hardly be described as thinking, even mistakenly, that I am sitting on a pile of hay; all I am doing is to use the *words* 'pile of hay'.

It might be thought that the two cases were not parallel, for while the meaning of 'pile of hay' is given by the characteristics which piles of hay must possess, the meaning of 'rude' is given by the attitude it expresses. The answer is that if 'thinking a thing rude' is to be described as having a particular attitude to it, then having an attitude presupposes, in this case, believing that certain conditions are fulfilled. If 'attitudes' were solely a matter of reactions such as wrinkling the nose, and tendencies to such things as making resolutions and scolding, then thinking something rude would not be describable solely in terms of attitudes. Either thinking something rude is not to be described in terms of attitudes, or attitudes are not to be described in terms of such things. Even if we could suppose that a particular individual could react towards conventional behaviour, or to walking slowly up to an English front door, *exactly* as most people react to behaviour which gives offence, this would not mean that he was to be described as thinking these things rude. And in any case the supposition is nonsense. Although he could behave in some ways as if he thought them rude, e.g. by scolding conventional or slow-walking children, but not turning daughters with these proclivities out of doors, his behaviour could not be just as if he thought them rude. For as the social reaction to conventional behaviour is not the same as the social reaction to offensive behaviour, he could not act in just the same way. He could not for instance apologize for what he would call his 'rudeness', for he would have to admit that it had caused no offence.

I conclude that whether a man is speaking of behaviour as rude or not rude, he must use the same criteria as anyone else, and that since the criteria are satisfied if O is true, it is impossible for him to assert O while denying R. It follows that if it is a sufficient condition of P's entailing Q that the assertion of P is inconsistent with the denial of Q, we have here an example of a non-evaluative premise from which an evaluative conclusion can be deduced.

It is of course possible to admit O while refusing to assert R, and this will not be like the refusal to say about prunes what one has already admitted about dried plums. Calling an action 'rude' is using a concept which a man might want to reject, rejecting the whole practice of praising and blaming embodied in terms such as 'polite' and 'rude'. Such a man would refuse to discuss points of etiquette, and arguments with him about what is rude would not so much break down as never begin. But once he did accept the question 'Is this rude?', he would have to abide by the rules of this kind of argument; he could not bring forward any evidence he liked, and he could not deny the relevance of any piece of evidence brought forward by his opponent. Nor could he say that he was unable to move from O to R on this occasion because the belief in O had not induced in him feelings or attitudes warranting the assertion of R. If he had agreed to discuss rudeness he had committed himself to accepting O as evidence for R, and evidence is not a sort of medicine which is taken in the hope that it will work. To suggest that he could refuse to admit that certain behaviour was rude because the right psychological state had not been induced, is as odd as to suppose that one might refuse to speak of the world as round because in spite of the good evidence of roundness a feeling of confidence in the proposition had not been produced. When given good evidence it is one's business to act on it, not to hang around waiting for the right state of mind. It follows that if a man is prepared to discuss questions of rudeness, and hence to accept as evidence the fact that behaviour causes a certain kind of offence, he cannot refuse to admit R when O has been proved.

The point of considering this example was to show that there may be the strictest rules of evidence even where an evaluative conclusion is concerned. Applying this principle to the case of moral judgements, we see that – for all that the non-naturalist has proved to the contrary – Bentham, for instance, may be right in saying that when used in conjunction with the principle of utility 'the words *ought* and *right* and *wrong*, and others of that stamp, have a meaning: when otherwise they have none'.[1] Anyone who uses moral terms at all, whether to assert or deny a moral proposition, must abide by the rules for their use, including the rules about what shall count as evidence for or against the moral judgement concerned. For anything that has yet been shown to the contrary these rules could be entailment rules, forbidding the assertion of factual propositions in conjunction with the denial of moral propositions. The only recourse of the man who refused to accept the things which counted in favour of a moral proposition as giving him a reason to do certain things

[1] *Principles of Morals and Legislation*, chap. I, x.

or to take up a particular attitude, would be to leave the moral discussion and abjure altogether the use of moral terms.

To say what Bentham said is not, then, to commit any sort of 'naturalistic fallacy'. It is open to us to enquire whether moral terms do lose their meaning when divorced from the pleasure principle, or from some other set of criteria, as the word 'rude' loses its meaning when the criterion of offensiveness is dropped. To me it seems that this is clearly the case; I do not know what could be meant by saying that it was someone's duty to do something unless there was an attempt to show why it mattered if this sort of thing was not done. How can questions such as 'what does it matter?', 'what harm does it do?', 'what advantage is there in . . . ?', 'why is it important?', be set aside here? Is it even to be suggested that the harm done by a certain trait of character could be taken, by some extreme moral eccentric, to be just what made it a virtue? I suggest that such a man would not even be a moral eccentric, any more than the man who used the word 'rude' of conventional behaviour was putting forward strange views about what was rude. Both descriptions have their proper application, but it is not here. How exactly the concepts of harm, advantage, benefit, importance, etc., are related to the different moral concepts, such as rightness, obligation, goodness, duty and virtue, is something that needs the most patient investigation, but that they are so related seems undeniable, and it follows that a man cannot make his own personal decision about the considerations which are to count as evidence in morals.

Perhaps it will be argued that this kind of freedom of choice is not ruled out after all, because a man has to decide for himself what is to count as advantage, benefit, or harm. But is this really plausible? Consider the man described by Hare as thinking that torturing is morally permissible.[1] Apparently he is not supposed to be arguing that in spite of everything torture is justifiable as a means of extracting confessions from enemies of the state, for the argument is supposed to be at an end when he has said that torturing people is permissible, and his opponent has said that it is not. How is he supposed to have answered the objection that to inflict torture is to do harm? If he is supposed to have said that

[1] 'Universalizability', *P.A.S.*, 1954–5, p. 304.

pain is good for a man in the long run, rather than bad, he will have to show the benefits involved, and he can no more choose what shall count as a benefit than he could have chosen what counted as harm. Is he supposed perhaps to count as harm only harm to himself? In this case he is guilty of *ignoratio elenchi*. By refusing to count as harm anything except harm to himself, he puts himself outside the pale of moral discussion, and should have explained that this was his position. One might compare his case to that of a man who in some discussion of common policy says 'this will be the best thing to do', and announces afterwards that *he* meant best for himself. This is not what the word 'best' does mean in the context of such a discussion.

It may be objected that these considerations about the evidence which must be brought for saying that one thing is good and another bad could not in any case be of the least importance; such rules of evidence, even if they exist, only reflecting the connection between our existing moral code and our existing moral terms; if there are no 'free' moral terms in our language, it can always be supposed that some have been invented – as indeed they will have to be invented if we are to be able to argue with people who subscribe to a moral code entirely different from our own. This objection rests on a doubtful assumption about the concept of *morality*. It assumes that even if there are rules about the grounds on which actions can be called good, right, or obligatory, there are no rules about the grounds on which a principle which is to be called a moral principle may be asserted. Those who believe this must think it possible to identify an element of feeling or attitude which carries the meaning of the word 'moral'. It must be supposed, for instance, that if we describe a man as being for or against certain actions, bringing them under universal rules, adopting these rules for himself, and thinking himself bound to urge them on others, we shall be able to identify him as holding moral principles, whatever the content of the principle at which he stops. But why should it be supposed that the concept of morality is to be caught in this particular kind of net? The consequences of such an assumption are very hard to stomach; for it follows that a rule which was admitted by those who obeyed it to be completely pointless could yet be recognized as a moral rule. If people happened to insist that no one should run round trees left handed, or look at hedgehogs in the light of the moon, this might count as a basic moral principle about which nothing more need be said.

I think that the main reason why this view is so often held in spite of these difficulties, is that we fear the charge of making a verbal decision in favour of our own moral code. But those who bring that charge are merely begging the question against arguments such as those given above. Of course if the rules we are refusing to call moral rules can really be given this name, then we are merely legislating against alien *moral codes*. But the suggestion which has been put forward is that this could not be the right description for rules of behaviour for which an entirely different defence is offered from that which we offer for our moral beliefs. If this suggestion is right, the difference between ourselves and the people who have these rules is not to be described as a difference of moral outlook, but rather as a difference between a moral and a non-moral point of view. The example of etiquette is again useful here. No one is tempted to say that the ruling out, *a priori*, of rules of etiquette which each man decides on for himself when he feels so inclined, represents a mere verbal decision in favour of our kind of socially determined standards of etiquette. On what grounds could one call a rule which someone was allowed to invent for himself a rule of *etiquette*? It is not just a fact about the use of our words 'rude', 'not done', etc., that they could not be applied in such a case; it is also a fact about etiquette that if terms in another language did appear in such situations they would not be terms of etiquette. We can make a similar point about the terms 'legal' and 'illegal' and the concept of law. If any individual was allowed to apply a certain pair of terms expressing approval and disapproval off his own bat, without taking notice of any recognized authority, such terms could not be legal terms. Similarly it is a fact about etiquette and law that they are both conventional as morality is not.

It may be that in attempting to state the rules which govern the assertion of moral propositions we shall legislate against a moral system radically opposed to our own. But this is only to say that we may make a mistake. The remedy is to look more carefully at the rules of evidence, not to assume that there cannot be any at all. If a moral system such as Nietzsche's has been refused recognition as a moral system, then we have got the criteria wrong. The fact that Nietzsche was a moralist cannot, however, be quoted in favour of the private enterprise theory of moral criteria. Admittedly Nietzsche said 'You want to decrease suffering; I want precisely to increase it' but he did not *just* say this. Nor did he offer as a justification the fact that suffering causes a tendency to absentmindedness, or lines on the human face. We recognize Nietzsche as a moralist because he tries to justify an increase in suffering by connecting it with strength as opposed to weakness, and individuality as opposed to conformity. That strength is a good thing can only be denied by someone who can show that the strong man overreaches himself, or in some other way brings harm to himself or other people. That individuality is a good thing is something that has to be shown, but in a vague way we connect it with originality, and with courage, and hence there is no difficulty in conceiving Nietzsche as a moralist when he appeals to such a thing.

In conclusion it is worth remarking that moral arguments break down more often than philosophers tend to think, but that the breakdown is of a different kind. When people argue about what is right, good, or obligatory, or whether a certain character trait is or is not a virtue, they do not confine their remarks to the adducing of facts which can be established by simple observation, or by some clear-cut technique. What is said may well be subtle or profound, and in this sort of discussion as in others, in the field of literary criticism for instance, or the discussion of character, much depends on experience and imagination. It is quite common for one man to be unable to see what the other is getting at, and this sort of misunderstanding will not always be resolvable by anything which could be called argument in the ordinary sense.

II

The moral point of view

KURT BAIER

Throughout the history of philosophy, by far the most popular candidate for the position of the moral point of view has been self-interest. There are obvious parallels between these two standpoints. Both aim at the good. Both are rational. Both involve deliberation, the surveying and weighing of reasons. The adoption of either yields statements containing the word 'ought'. Both involve the notion of self-mastery and control over the desires. It is, moreover, plausible to hold that a person could not have a reason for doing anything whatsoever unless his behavior was designed to promote his own good. Hence, if morality is to have the support of reason, moral reasons must be self-interested, hence the point of view of morality and self-interest must be the same. On the other hand, it seems equally obvious that morality and self-interest are very frequently opposed. Morality often requires us to refrain from doing what self-interest recommends or to do what self-interest forbids. Hence morality and self-interest cannot be the same points of view.

I. SELF-INTEREST AND MORALITY

Can we save the doctrine that the moral point of view is that of self-interest? One way of circumventing the difficulty just mentioned is to draw a distinction between two senses of 'self-

---

Reprinted from The Moral Point of View *(Cornell University Press, 1958), pp. 187–213, by permission of the author and publishers. There is a discussion of some of the points raised here in R.N.Bronaugh, 'Formal Criteria for Moral Rules',* Mind, *1968.*

interest', shortsighted and enlightened. The shortsighted egoist always follows his short-range interest without taking into consideration how this will affect others and how their reactions will affect him. The enlightened egoist, on the other hand, knows that he cannot get the most out of life unless he pays attention to the needs of others on whose good will he depends. On this view, the standpoint of (immoral) egoism differs from that of morality in that it fails to consider the interests of others even when this costs little or nothing or when the long-range benefits to oneself are likely to be greater than the short-range sacrifices.

This view can be made more plausible still if we distinguish between those egoists who consider each course of action on its own merits and those who, for convenience, adopt certain rules of thumb which they have found will promote their long-range interest. Slogans such as 'Honesty is the best policy', 'Give to charity rather than to the Department of Internal Revenue', 'Always give a penny to a beggar when you are likely to be watched by your acquaintances', 'Treat your servants kindly and they will work for you like slaves', 'Never be arrogant to anyone – you may need his services one day', are maxims of this sort. They embody the 'wisdom' of a given society. The enlightened long-range egoist may adopt these as rules of thumb, that is, as prima facie maxims, as rules which he will observe unless he has good evidence that departing from them will pay him better than abiding by them. It is obvious that the rules of behavior adopted by the enlightened egoist will be very similar to those of a man who rigidly follows our own moral code.

Sidgwick appears to believe that egoism is one of the legitimate 'methods of ethics', although he himself rejects it on the basis of an 'intuition' that it is false. He supports the legitimacy of egoism by the argument that everyone could consistently adopt the egoistic point of view. 'I quite admit that when the painful necessity comes for another man to choose between his own happiness and the general happiness, he must as a reasonable being prefer his own, i.e. it is right for him to do this on my principle.'[1] The consistent enlightened egoist satisfies the

[1] Henry Sidgwick, *The Methods of Ethics*, 7th edn. (London: Macmillan and Co., 1907), pref. to the 6th edn., p. xvii.

categorical imperative, or at least one version of it, 'Act only on that maxim whereby thou canst at the same time will that it should become a universal law.'

However, no 'intuition' is required to see that this is not the point of view of morality, even though it can be universally adopted without self-contradiction. In the first place, a consistent egoist adopts for all occasions the principle 'everyone for himself' which we allow (at most) only in conditions of chaos, when the normal moral order breaks down. Its adoption marks the return to the law of the jungle, the state of nature, in which the 'softer', 'more chivalrous' ways of morality have no place.

This point can be made more strictly. It can be shown that those who adopt consistent egoism cannot make moral judgments. Moral talk is impossible for consistent egoists. But this amounts to a *reductio ad absurdum* of consistent egoism.

Let B and K be candidates for the presidency of a certain country and let it be granted that it is in the interest of either to be elected, but that only one can succeed. It would then be in the interest of B but against the interest of K if B were elected, and vice versa, and therefore in the interest of B but against the interest of K if K were liquidated, and vice versa. But from this it would follow that B ought to liquidate K, that it is wrong for B not to do so, that B has not 'done his duty' until he has liquidated K; and vice versa. Similarly K, knowing that his own liquidation is in the interest of B and therefore anticipating B's attempts to secure it, ought to take steps to foil B's endeavors. It would be wrong for him not to do so. He would 'not have done his duty' until he had made sure of stopping B. It follows that if K prevents B from liquidating him, his act must be said to be both wrong and not wrong – wrong because it is the prevention of what B ought to do, his duty, and wrong for B not to do it; not wrong because it is what K ought to do, his duty, and wrong for K not to do it. But one and the same act (logically) cannot be both morally wrong and not morally wrong. Hence in cases like these morality does not apply.

This is obviously absurd. For morality is designed to apply in just such cases, namely, those where interests conflict. But if the point of view of morality were that of self-interest, then there could *never* be moral solutions of conflicts of interest. However, when there are conflicts of interest, we always look for a 'higher point of view, one from which such conflicts can be settled. Consistent egoism makes everyone's private interest the 'highest court of appeal'. But by 'the moral point of view' we *mean* a point of view which is a court of appeal for conflicts of interest. Hence it cannot (logically) be identical with the point of view of self-interest. Sidgwick is, therefore, wrong in thinking that consistent egoism is one of the 'legitimate methods of ethics'. He is wrong in thinking that an 'intuition' is required to see that it is not the correct moral point of view. That it is not can be seen in the same way in which we can 'see' that the Court of Petty Sessions is not the Supreme Court.

II. MORALITY INVOLVES DOING THINGS ON PRINCIPLE

Another feature of consistent egoism is that the rules by which a consistent egoist abides are merely rules of thumb. A consistent egoist has only one supreme principle, to do whatever is necessary for the realization of his one aim, the promotion of his interest. He does not have *principles*, he has only an aim. If one has adopted the moral point of view, then one acts on principle and not merely on rules of thumb designed to promote one's aim. This involves conforming to the rules whether or not doing so favors one's own or anyone else's aim.

Kant grasped this point even if only obscurely. He saw that adopting the moral point of view involves acting on principle. It involves conforming to rules even when doing so is unpleasant, painful, costly, or ruinous to oneself. Kant, furthermore, argued rightly that, since moral action is action on principle (and not merely in accordance with rules of thumb), a moral agent ought not to make exceptions in his own favor, and he interpreted this to mean that moral rules are absolutely inflexible and without exceptions. Accordingly he concluded that if 'Thou shalt not kill' states a moral rule, then any and every act correctly describable as an act of killing someone must be said to be morally wrong.

Kant also saw that this view required him to reject some of our deepest moral convictions; we certainly think that the killing of a man in self-defense or by the hangman is not morally wrong. Kant was prepared to say that our moral convictions are wrong on this point. Can we salvage these moral convictions? The only alternative, to say that acting on principle does not require us not to make exceptions in our own favor, seems to be equally untenable.

It is therefore not surprising that many philosophers have abandoned Kant's (and the commonsense) view that the moral rightness of an act is its property of being in accordance with a moral rule or principle. Thus, the deontologists claim that rightness is a simple property which we can 'see' or 'intuit' in an act, and the utilitarians, that rightness is a complex property, namely, the tendency of an act to promote the greatest happiness of the greatest number. But, as is well known, these accounts are not plausible and lead to considerable difficulties.

However, this whole problem arises only because of a confusion, the confusion of the expression 'making an exception to a rule' with the expression 'a rule has an exception'. As soon as this muddle is cleared away, it can be seen that Kant is right in saying that acting on principle implies making no exception in anyone's favor, but wrong in thinking that therefore all moral rules must be absolutely without exception.

'No parking in the city' has a number of recognized exceptions which are part of the rule itself, for example, 'except in the official parking areas', 'except in front of a parking meter', 'except on Saturday mornings and after 8 p.m. every day'. A person who does not know the recognized exceptions does not completely know the rule, for these exceptions more precisely define its range of application. A policeman who is not booking a motorist parking in front of a parking meter is not granting exemption to (making an exception in favor of) this motorist. On the contrary, he is administering the rule correctly. If he did apply the no-parking rule to the motorist, *he* would be applying it where *it* does not apply, because this is one of the recognized exceptions which are *part* of the rule. On the other hand, a policeman who does not book a motorist parking his vehicle in a prohibited area at peak hour on a busy day is making an exception in the motorist's favor. If he does so because the man is his friend, he illegitimately grants an exemption. If he does so because the motorist is a doctor who has been called to attend to a man lying unconscious on the pavement, this is a 'deserving case' and he grants the exemption legitimately.

Apply this distinction to the rules of a given morality. Notice first that moral rules differ from laws and regulations in that they are not administered by special administrative organs such as policemen and magistrates. Everyone 'administers' them himself. Nevertheless, it makes sense to speak of making exceptions in one's own favor. For one may refuse to apply the rule to oneself when one knows that it does apply, that is to say, one may refuse to observe it even when one knows one should. And what is true of making exceptions in one's own favor is true also of making them in favor of someone else. It is almost as immoral to make exceptions in favor of one's wife, son, or nephew as in favor of oneself.

When we say, therefore, that a person who has killed a burglar in self-defense has not done anything wrong, we are not making an exception in the houseowner's favor. It is much nearer the truth to say that, in our morality, the rule 'Thou shalt not kill' *has several recognized exceptions*, among them 'in self-defense'. We can say that a man does not know fully our moral rule 'Thou shalt not kill' if he does not know that it has, among others, this exception.

Like other rules of reason, our moral convictions are so only *presumptively*. Killing is wrong *unless* it is killing in self-defense, killing by the hangman, killing of an enemy in wartime, accidental killing, and possibly mercy killing. If it is one of these types of killing, then it is *not* wrong.

Even if it is one of the wrongful acts of killing, it is so only prima facie, other things being equal. For there may have been an overriding moral reason in favor of killing the man, for example, that he is about to blow up a train and that this is the only way of stopping him.

One further point should be made to avoid misunderstanding. Unlike laws and regulations, moral rules have not been laid down by anyone. Knowing moral rules cannot, therefore, involve knowing exactly what a certain person has enjoined and forbidden and what exceptions he has allowed, because there is no such person. In the case of regulations and laws, it was precisely this knowledge which enabled us to draw the distinction between saying that someone was granting an exception and saying that he was merely applying the rule which, for cases of this sort, provided for an exception. Our distinction seems to collapse for moral rules.

However, the answer to this is simple. When a magistrate is empowered to make exceptions or grant exemptions in 'deserving cases', the question of what is a 'deserving case' is not of course answered in the regulation itself. If it were, the magistrate would not be exercising his power to grant exemption, but would simply apply the regulation as provided in it. How, then, does the magistrate or policeman know what is a deserving case? The doctor who parks his car in a prohibited spot in order to attend to an injured man is such a case, namely, a *morally deserving case*. The principles in accordance with which policemen or magistrates grant exemptions to existing regulations are moral principles. In the case of moral rules, there cannot be any distinction between exceptions which are part of the rule and deserving cases. *Only* deserving cases can be part of the moral rule, and *every* deserving case is properly part of it. Hence while in the case of laws and regulations there is a reason for going beyond the exceptions allowed in the regulation itself (when there is a morally deserving instance), in the case of moral rules there is no such reason. For all deserving cases are, from the nature of the case, part of the moral rule itself. Hence it is never right to make an exception to a moral rule in anyone's favor. Kant is therefore quite right in saying that it is always wrong to make exceptions to moral rules in one's own favor (and for that matter in anyone else's), but he is wrong in thinking that this makes moral rules inflexible.

All this follows from the very nature of moral principles. They are binding on everyone alike quite irrespective of what are the goals or purposes of the person in question. Hence self-interest cannot be the moral point of view, for it sets every individual one supreme goal, his own interest, which overrules all his other maxims.

III. MORAL RULES ARE MEANT FOR EVERYBODY

The point of view of morality is inadequately characterized by saying that *I* have adopted it if *I* act on principles, that is, on rules to which I do not make exceptions whenever acting on them would frustrate one or the other of my purposes or desires. It is characterized by greater universality than that. It must be thought of as a standpoint from which principles are considered as being acted on *by everyone*. Moral principles are not merely principles, on which a person must always act without making exceptions but they are principles *meant for everybody*.

It follows from this that the teaching of morality must be completely universal and open. Morality is meant to be taught to all members of the group in such a way that everyone can and ought always to act in accordance with these rules. It is not the preserve of an oppressed or privileged class or individual. People are neglecting their duties if they do not teach the moral rules to their children. Children are removed from the homes of criminals because they are not likely to be taught the moral rules there. Furthermore, moral rules must be taught quite openly and to everybody without discrimination. An esoteric code, a set of precepts known only to the initiated and perhaps jealously concealed from outsiders, can at best be a religion, not a morality. 'Thou shalt not eat beans and this is a secret' or 'Always leave the third button of your waistcoat undone, but don't tell anyone except the initiated members' may be part of an esoteric religion, but not of a morality. 'Thou shalt not kill, but it is a strict secret' is absurd. 'Esoteric morality' is a contradiction in terms. It is no accident that the so-called higher religions were imbued with the missionary spirit, for they combine the beliefs of daemons and gods and spirits characteristic of primitive religions with *a system of morality*. Primitive religions are not usually concerned to proselytize. On the contrary, they are imbued with the spirit of the exclusive trade secret. If one thinks of one's religion as concentrated wisdom of life revealed solely to the *chosen* people, one will regard it as the exclusive property of the club, to be confined to the elect. If, on the other hand, the rules are thought to be for everyone, one must in consistency want to spread the message.

The condition of universal teachability yields three other criteria of moral rules. They must not, in the first place, be 'self-frustrating'. They are so if their purpose is frustrated as soon as everybody acts on them, if they have a point only when a good many people act on the opposite principle. Someone might, for instance, act on the maxim 'When you are in need, ask for help, but never help another man when he is in need'. If everybody adopted this principle, then their adoption of the second half would frustrate what obviously is the point of the adoption of the first half, namely, to get help when one is in need. Although such a principle is not self-contradictory – for anybody could consistently adopt it – it is nevertheless objectionable from the moral point of view, for it could not be taught openly to everyone. It would then lose its point. It is a parasitic principle, useful to anyone only if many people act on its opposite.

The same is true of 'self-defeating' and 'morally impossible' rules. A principle is self-defeating if its point is defeated as soon as a person lets it be known that he has adopted it, for example, the principle 'Give a promise even when you know or think that you can never keep it, or when you don't intend to keep it'. The very point of giving promises is to reassure and furnish a guarantee to the promisee. Hence any remark that throws doubt on the sincerity of the promiser will defeat the purpose of making a promise. And clearly to *let it be known* that one gives promises even when one knows or thinks one cannot, or when one does not intend to keep them, is to raise such doubts. And to say that one acts on the above principle is to imply that one may well give promises in these cases. Hence to reveal that one acts on this principle will tend to defeat one's own purpose.

It has already been said that moral rules must be capable of being taught openly, but this rule is self-defeating when taught openly, for then everyone would be known to act on it. Hence it cannot belong to the morality of any group.

Lastly, there are some rules which it is literally impossible to teach in the way the moral rules of a group must be capable of being taught, for example, the rule 'Always assert what you think not to be the case'. Such *morally impossible* rules differ from self-frustrating and self-defeating rules in that the latter could have been taught in this way, although it would have been quite senseless to do so, whereas the former literally cannot be so taught. The reason why the above rule cannot be taught in this way is that the only possible case of acting on it, doing so secretly, is ruled out by the conditions of *moral teaching*.

(1) Consider first someone secretly adopting this rule. His remarks will almost always mislead people, for *he will be taken to be saying what he thinks true*, whereas he *is* saying the opposite. Moreover, in most cases what he thinks (and not what he says) will be true. Thus, it will usually be the case that *p* is true when he says 'not-*p*', and not-*p* when he says '*p*', whereas people will take it that *p* is true when he says '*p*', and not-*p* when he says 'not-*p*'. Thus communication between him and other people breaks down, since they will almost always be misled by him whether he wishes to mislead them or not. The possibility of communication depends on a speaker's ability *at will* to say either what he thinks to be the case or what he thinks not to be the case. Our speaker cannot communicate because by his principle he is forced to mislead his hearers.

Thus, anyone secretly adopting the principle 'Always assert what you think not to be the case' cannot communicate with others since he is bound to mislead them whether he wants to or not. Hence he cannot possibly teach the principle to anybody. And if he were to teach the principle without having adopted it himself, then, although he would be understood, those who adopted it would not. At any rate, since moral teaching involves teaching rules such as the taught may openly avow to be observing, this case is ruled out. A principle which is taught for secret acceptance only cannot be embodied in a *moral* rule of the group.

(2) Of course, people might soon come to realize what is the matter with our man. They may discover that in order not to be misled by what he says they have only to substitute '*p*' for 'not-*p*' and vice versa. But if they do this, then they have interpreted his way of speaking, not as a reversal of the general presumption

that one says what one thinks is the case (and not the opposite) but as a change of the use of 'not'. In his language, it will be said, 'not' has become an affirmation sign, negation being effected by omitting it. Thus, if communication is to be possible, we must interpret as a change in usage what is intended as the reversal of the presumption that every assertion conveys what the assertor believes to be the case.

If everyone were, by accident, to adopt simultaneously and secretly our principle 'Always assert what you think is not the case', then, for some time at least, communication would be impossible. If, on the other hand, it were adopted openly, then communication would be possible, but only if the adoption of this principle were to be accompanied by a change in the use of 'not' which would completely cancel the effect of the adoption of the principle. In that case, however, it can hardly be said that the principle has been adopted.

(3) The case we are considering is neither (1) nor (2). We are considering the open teaching of the principle 'Always assert what you think is not the case', for open acceptance by everybody, an acceptance which is not to be interpreted as a change in the use of 'not'. But this is nonsense. We cannot all *openly* tell one another that we are always going to mislead one another in a certain way and insist that we must continue to be misled, though we know how we could avoid being misled. I conclude that this principle could not be embodied in a rule belonging to the morality of any group.

These points are of general interest in that they clarify some valuable remarks contained in Kant's doctrine of the categorical imperative. In particular they clarify the expression 'can will' contained in the formulation 'Act so that thou *canst will* thy maxim to become a universal law of nature'. 'Canst will' in one sense means what I have called 'morally possible'. Your maxim must be a formula which is morally possible, that is, which is logically capable of being a rule belonging to the morality of some group, as the maxim 'Always lie' is not. No one *can* wish that maxim to be a rule *of some morality*. To say that one is wishing it is to contradict oneself. One cannot wish it any more than one can wish that time should move backwards.

The second sense of 'can will' is that in which no rational person can will certain things. Self-frustrating and self-defeating moral rules are not morally impossible, they are merely senseless. No rational person could wish such rules to become part of any morality. That is to say, anyone wishing that they should would thereby expose himself to the charge of irrationality, like the person who wishes that he should never attain his ends or that he should (for no reason at all) be plagued by rheumatic pains throughout his life.

The points just made also show the weakness of Kant's doctrine. For while it is true that someone who acts on the maxim 'Always lie' acts on a morally impossible one, it is not true that every liar necessarily acts on that maxim. If he acts on a principle at all, it may, for instance, be 'Lie when it is the only way to avoid harming someone', or 'Lie when it is helpful to you and harmful to no one else', or 'Lie when it is entertaining and harmless'. Maxims such as these can, of course, be willed in either of the senses explained.

IV. MORAL RULES MUST BE FOR THE GOOD OF EVERYONE ALIKE

The conditions so far mentioned are merely formal. They exclude certain sorts of rule as not coming up to the formal requirements. But moral rules should also have a certain sort of content. Observation of these rules should be *for the good of everyone alike.* Thrasymachus' view that justice is the advantage of the stronger, if true of the societies of his day, is an indictment of their legal systems from the moral point of view. It shows that what goes by the name of morality in these societies is no more than a set of rules and laws which enrich the ruling class at the expense of the masses. But this is wrong because unjust, however much the rules satisfy the formal criteria. For given certain initial social conditions, formal equality before the law may favour certain groups and exploit others.

There is one obvious way in which a rule may be for the good of everyone alike, namely, if it furthers the common good. When I am promoted and my salary is raised, this is to my advantage. It will also be to the advantage of my wife and my family and possibly of a few other people – it will not be to the advantage of my colleague who had hoped for promotion but is now excluded. It may even be to his detriment if his reputation suffers as a result. If the coal miners obtain an increase in their wages, then this is to the advantage of coal miners. It is for their common good. But it may not be to the advantage of anyone else. On the other hand, if production is raised and with it everyone's living standard, that is literally to everyone's advantage. The rule 'Work harder', if it has these consequences, is for the common good of all.

Very few rules, if any, will be for the common good of everyone. But a rule may be in the interest of everyone alike, even though the results of the observation of the rule are not for the common good in the sense explained. Rules such as 'Thou shalt not kill', 'Thou shalt not be cruel', 'Thou shalt not lie' are obviously, in some other sense, for the good of everyone alike. What is this sense? It becomes clear if we look at these rules from the moral point of view, that is, that of an independent, unbiased, impartial, objective, dispassionate, disinterested observer. Taking such a God's-eye point of view, we can see that it is in the interest of everyone alike that everyone should abide by the rule 'Thou shalt not kill'. From the moral point of view, it is clear that it is in the interest of everyone alike if everyone alike should be allowed to pursue his own interest provided this does not adversely affect someone else's interests. Killing someone in the pursuit of my interests would interfere with his.

There can be no doubt that such a God's-eye point of view is involved in the moral standpoint. The most elementary teaching is based on it. The negative version of the so-called Golden Rule sums it up: 'Don't do unto others as you would not have them do unto you'. When we teach children the moral point of view, we try to explain it to them by getting them to put themselves in another person's place: 'How would you like to have that done to you!' 'Don't do evil', the most readily accepted moral rule of all, is simply the most general form of stating this prohibition. For doing evil is the opposite of doing good. Doing good is doing for another person what, if he were following (self-interested) reason, he would do for himself. Doing evil is doing to another person what it would be contrary to reason for him to do to himself. Harming another, hurting another, doing to another what he dislikes having done to him are the specific forms this takes. Killing, cruelty, inflicting pain, maiming, torturing, deceiving, cheating, rape, adultery are instances of this sort of behavior. They all violate the condition of 'reversibility', that is, that the behavior in question must be acceptable to a person whether he is at the 'giving' or 'receiving' end of it.

It is important to see just what is established by this condition of being for the good of everyone alike. In the first place, anyone is doing wrong who engages in nonreversible behavior. It is irrelevant whether he knows that it is wrong or not, whether the morality of his group recognizes it or not. Such behavior is 'wrong in itself', irrespective of individual or social recognition, irrespective of the consequences it has. Moreover, every single act of such behavior is wrong. We need not consider the whole group or the whole of humanity engaging in this sort of behavior, but only a single case. Hence we can say that all nonreversible behavior is morally wrong; hence that anyone engaging in it is doing what, prima facie, he ought not to do. We need not consider whether this sort of behavior has harmful consequences, whether it is forbidden by the morality of the man's group, or whether he himself thinks it wrong.

The principle of reversibility does not merely impose certain prohibitions on a moral agent, but also certain positive injunctions. It is, for instance, wrong – an omission – not to help another person when he is in need and when we are in a position to help him. The story of the Good Samaritan makes this point. The positive version of the Golden Rule makes the same point more generally: 'Do unto others as you would have them do unto you'. Note that it is wrong – not merely not meritorious – to omit to help others when they are in need and when you are in a position to help them. It does not follow from this, however, that it is wrong not to promote the greatest good of the greatest number, or not to promote the greatest amount of good in the world. Deontologists and utilitarians alike make the mistake of thinking that it is one, or the only one, of our moral duties to 'do the optimific act'. Nothing could be further from the truth. We do not have a duty to do good to others or to ourselves, or to others and/or to ourselves in a judicious mixture such that it produces the greatest possible amount of good in the world. We are morally required to do good only to those who are actually in need of our assistance. The view that we always ought to do the optimific act, or whenever we have no more stringent duty to perform, would have the absurd result that we are doing wrong whenever we are relaxing, since on those occasions there will always be opportunities to produce greater good than we can by relaxing. For the relief of suffering is always a greater good than mere enjoyment. Yet it is quite plain that the worker who, after a tiring day, puts on his slippers and listens to the wireless is not doing anything he ought not to, is not neglecting any of his duties, even though it may be perfectly true that there are things he might do which produce more good in the world, even for himself, than merely relaxing by the fireside.

V. UPSETTING AND RESTORING THE MORAL EQUILIBRIUM

So far, we have considered only primary moral rules, that is, those which prohibit or enjoin certain types of behavior, such as 'Thou shalt not kill', 'Thou shalt not steal', 'Thou shalt help thy neighbor when he is in need of your help', and so on. Secondary rules of morality are those which prohibit or enjoin certain types of behavior *in response to* some 'upset of the moral balance', for example, 'An eye for an eye, a tooth for a tooth', 'Let him who is free from guilt throw the first stone', 'One good turn deserves another'.

What is it to 'upset the moral balance'? The moral balance is preserved when everyone is 'strictly minding his own business'. Plato was right in connecting morality with minding one's own business; he was wrong only in his explanation of the connection. Minding one's own business and not interfering with anyone else are not all there is to morality, though it is true that when everyone minds his own business the moral equilibrium is maintained. This equilibrium can be upset in two quite different ways. I may behave in a manner which upsets the moral balance against me or in my favor. I may accumulate a moral debit or credit account;

the first when I do what I ought not to do, the second when I do 'more than my duty'; the first when I violate a 'rule of duty', the second when I observe a 'rule of supererogation'. When, for example, I kill someone, steal something, am cruel to someone, or commit adultery, I am accumulating a moral debit balance. If, on the other hand, at great risk to myself I save someone's life or make great financial sacrifices for the sake of a good cause, I am acquiring a moral credit balance. It is for cases of this sort that the secondary moral rules are devised. Primary moral rules define what it is, morally speaking, to mind one's own business, to preserve the moral equilibrium. Secondary moral rules indicate what is to be done by whom when the balance has been upset.

Secondary moral rules are determined by the concept of desert, of positive or negative moral merit. They state what a person deserves, that is, ought to get or have done to him, as a result of the upset of the moral balance. A person who has not upset the moral balance deserves nothing. He has neither positive nor negative moral merit.

The aim of a morality is to prevent the upsetting of the moral equilibrium by violation of 'rules of duty' and to encourage it by the observation of 'rules of supererogation'. At the same time, the methods of deterring and encouraging potential rule breakers must not themselves interfere with the primary rules. The secondary rules are therefore seen as designed to 'restore the moral balance'. They have the object of deterring or encouraging rule breakers, but also of bringing the process to an end. When the balance is 'restored', the secondary rules no longer apply.

Take first the case of preventing violations of rules of duty. An obvious, if crude, way of 'restoring the moral equilibrium' is provided by the institution of revenge. The person injured returns the harm. The supreme principle governing such secondary rules is 'One bad turn deserves another'. This has the serious disadvantage that it is difficult to 'restore' the moral equilibrium. Since revenge is itself the infliction of harm on an individual, the secondary rule applies again. In the institution of the vendetta or the blood feud, what is designed to discourage violations of primary moral rules in fact leads to endless mutual harming.

The substitution of punishment for revenge remedies this

drawback. The infliction of hardship on the wrongdoer is taken out of the hands of the injured person or his aggrieved relations and handed over to a disinterested official. By making a ceremony of it, it is clearly indicated that this is not intended as merely the infliction of harm on an individual, but as the application of a secondary moral rule designed to 'restore' the moral balance. The object of the practice is to deter future wrongdoers. The infliction of hardship on a given individual is justified by his prior violation of a primary moral rule. There is now no aggrieved person left. Punishment has restored the moral equilibrium. The wrongdoer has expiated, atoned for, his wrong. Everyone has a clean slate again. It is wrong for the aggrieved to continue to harbor a grudge, to refuse to forgive the wrongdoer.

The situation is somewhat different in the case of an upset of the moral balance by observing (not breaking) a rule of supererogation. Obviously, the point of these rules is that they should be observed rather than broken, although observing them (not breaking them) constitutes an upset of the moral equilibrium. Such breaches of the moral equilibrium are desirable. In order to encourage them, we have secondary rules of morality, guided by the general principle 'One good turn deserves another'. We say that a person who engages in works of supererogation thereby acquires moral desert or merit.

However, in the case of the personal requital of a good turn, there is nothing undesirable about the unending reciprocation of such good turns. It is desirable that the person who received the returns of gratitude should in turn feel grateful. There is, therefore, no reason why the state should take personal reciprocations out of the hands of the recipients and put communal rewards in their place. 'Mine is the vengeance' has a point; 'Mine is the gratitude' has not.

There are two conceptions which belong in this secondary field of morality, but which have usually been assigned to the primary field, obligation and justice. When we say that we are under an obligation to someone, we mean that we ought to restore the moral equilibrium by 'discharging our obligation'. To discharge one's obligations *is* to restore the moral equilibrium. Doing so terminates the special moral relationship created between two people by the upset of the moral balance which gave rise to the obligation. To say that killing is wrong or that one ought not to kill is to say something that does not involve the secondary field of morality. To say that one ought to discharge one's obligations is to say something that does. Many confusions and paradoxes could have been avoided if this distinction had been clearly grasped.

Justice is a moral concept involving the secondary field of morality. For 'doing justice' means 'giving to everyone what he deserves', that is, restoring the moral equilibrium by the appropriate action – giving a reward, repaying a debt, passing sentence, administering punishment, and the like. Plato's examination of morality in the Republic is vitiated by his failure to distinguish between the primary notions of rightness and wrongness and the secondary notion of justice.

VI. SOCIAL MORALITY

We have so far considered absolute morality only. As we have noted, the moral point of view is characterized by a formal and a material condition. The formal condition is this: a man cannot be said to have adopted the moral point of view unless he is prepared to treat the moral rules as principles rather than mere rules of thumb, that is, to do things on *principle* rather than merely to act purposively, merely to aim at a certain end. And, furthermore, he must act on rules which are meant for everybody, and not merely for himself or some favored group. The material condition is this: the rules must be for the good of everyone alike. This does not mean that they must be for the common good of all human beings, past, present, and future, for such a condition would be impossible to satisfy. Its meaning can be elucidated by setting forth the criteria of saying that a rule is for the good of everyone alike. As far as absolute morality is concerned, only one condition must be satisfied, namely, that these rules should be 'reversible', that is, not merely for the good of the agent, but at least not detrimental to the persons who are affected by the agent's behavior.

An examination of social conditions will yield some further

criteria of 'being for the good of everyone alike'. A society is more than just a number of individuals living in a certain area and behaving in ways directly affecting others, such as killing, maiming and robbing. Life in society involves a social framework which multiplies the points of contact between individuals and which can transform the effects of a man's behavior on his fellow men. Within a given social framework, behavior may be harmful which is not, from its nature, the infliction of harm on another. It may be harmful only if and because a great many people in that society engage in it. No harm is done if one person walks across the lawn. But the lawn is ruined if everyone does. No harm is done if one person uses the gas. But if everyone uses it during peak hours, then the gas supply may break down, and everyone will be adversely affected.

That such behavior is morally objectionable is widely recognized. We acknowledge that it is, by the well-known formula 'You can't do that; what if everyone did the same!' Kant thought of it as the core of his categorical imperative, 'Act only on that maxim whereby thou canst at the same time will that it should become a universal law.' This is precisely what we 'cannot will' in the cases in question. Although it is not true that, as Kant put it, a will willing such a maxim to become a universal law would, literally, contradict itself, nevertheless, in making such a maxim *a universal law*, one would enjoin people to do evil, and such a law would obviously be wrong.

It is, however, important to distinguish behavior which is 'nonuniversalizable' from behavior that is 'nonreversible'. The latter can be seen to be *wrong in itself*, irrespective of the consequences and of how many people engage in it. This is not so in the case of nonuniversalizable behavior. There we have to consider the consequences, and not merely of a single act but of a great many of them. There is nothing wrong in itself with putting one straw on the camel's back, but one of them will be the last.

What exactly does this prove? That no one is allowed to lay even one straw on the camel's back? That every act of this kind is wrong? Surely not. Before we can say that any act of this sort is wrong, a number of conditions must be satisfied.

In the first place, all concerned must be *equally entitled* to behave in the nonuniversalizable way. It would, for instance, be most undesirable if everyone had dinner at 6.30 p.m., for all the nation's service would then come to a standstill at that time. But it cannot follow from this that eating at 6.30 p.m. is wrong for everyone. It cannot follow because the argument applies equally for any time, and it must be all right to eat at *some* time. Of course, there is no serious problem here. Not everyone is equally entitled to have his dinner at 6.30 p.m. Those who are on duty at that time must have it before or after or while they are attending to their duties.

There are further conditions. If everyone were celibate all his life, mankind would die out, or, at any rate, the number would soon be so seriously reduced as to make life unbearable. Those who do not find the prospect of the end of the human race upsetting will have to admit that the return to primitive conditions is undesirable. Again, if everyone suddenly stopped smoking, drinking, gambling, and going to the pictures, some states might go bankrupt and this would be undesirable. All the same, it can hardly be true that abstinence in matters of sex, smoking, drinking, gambling, and visits to the cinema can be wrong in any and every case, even though we are surely all equally entitled to refrain from these ways of spending our time.

There must, therefore, be a further condition. Everyone must not only be equally entitled to engage in these forms of activity, but people must also be inclined to do so. There would have to be a real danger that, unless they are stopped somehow, many will engage in this sort of behavior. People are lazy, so they will not go to the polling booth or make the detour round the newly planted lawn. People like picking flowers, so they will destroy the rare wild flowers. People want to heat their rooms, so they will want to use their radiators during peak hours. But there is no great danger that they will all go celibate, or give up smoking and drinking.

This point, by the way, shows that non-universalizability cannot be adduced to show that suicide is wrong. Suicide is no more wrong than celibacy and for the same reason. People are less keen on suicide even than on celibacy. There is no danger of the race dying out. In fact, all over the world people are so keen on procreation that the suicide rate could go up a long way before anyone need be alarmed. Of course if, one day, life and sex were to become burdens to all of us and if, nevertheless, it really is desirable that the race should go on, then reckless suicide or slothful celibacy might become morally wrongful types of conduct. Until then, those weary of life and sex need not have a bad conscience about their uncommon indulgences.

There is one further point in this. To say that it is wrong to walk across the lawn or switch on the gas during peak hours, provided (a) it would have undesirable consequences *if* everyone did it, (b) we are all equally entitled to do it, and (c) doing it is an indulgence, not a sacrifice, amounts to saying that since refraining from doing these things is a sacrifice such a sacrifice for the common good should not be demanded of one or a few only, but equally of all, even if a universal sacrifice is not needed. Since no one is more entitled than anyone else to indulge himself and since *all* cannot do so without the undesirable consequences which no one wants, *no one* should be allowed to indulge himself.

Now the conditions are complete. If the behavior in question is such that (i) the consequences would be undesirable if everyone did it, (ii) all are equally entitled to engage in it, and (iii) engaging in this sort of behavior is an indulgence, not a sacrifice, then such behavior *should be prohibited by the morality of the group*.

But now suppose that it is not prohibited. Is it wrong all the same? Kant certainly thought so. I think he is mistaken. For since, by indulging in the behavior in question, I am not actually doing any harm, my behavior is not wrong in itself, but only when taken in conjunction with that of others. I cannot prevent the evil by refraining. Others must refrain too. In the case of nonreversible behavior, *my action alone* is the cause of the evil. I can avoid the evil if I refrain. In the case under discussion, however, if I have reason to suppose that the others will not refrain, I surely have reason not to refrain either, as my only reason for refraining is my desire to avoid causing the evil consequences. If these cannot be avoided, I have no reason not to indulge myself. If the grass is not going to grow anyway, why should I make the detour?

It is no good arguing that I am not entitled to do wrong just because other people might or probably would. For I am not doing wrong. I have no moral reason for the sacrifice. I need no justification or excuse, for my behavior is wrong only *if I have no reason to think* that others will refuse to make the sacrifice. If I have reason to think they will refuse to make it, then I have reason to think that my own sacrifice will be in vain; hence I have reason against making it.

Of course, if the results are *very* undesirable and my sacrifice is *very* small and I am not very certain what the others will do, I should take the risk of making the sacrifice even if it turns out to have been in vain. But, otherwise, reason will support the opposite course.

The situation is different if the morality or the custom or the law of the group does already contain a rule forbidding such behavior. If there is such a rule, then the behavior is wrong, for such a rule has the backing of morality. As we have said, a group ought to have rules forbidding nonuniversalizable behavior. And when there is such a rule, then the community has regulated behavior of this sort and I ought to do my share toward the success of the regulation.

I should like to add one word about the morality of individual initiative in these matters. Some people think that individuals should go ahead with a good example and not wait until the rule-making powers of the group are used. Others argue that this is putting too great a burden on the public-spirited. Thus, compulsory military service with exemptions granted to those engaged in important national industries is said by some to be fairer, volunteering for national service is said by others to be morally preferable. I can see no reason for the latter view. It may indeed seem preferable from the military point of view, for it may be argued that volunteers are better soldiers. But there is no reason why if keenness is wanted volunteers should not have preferential rights to serve in the army rather than in industry. On the other hand, there is no reason why the sacrifices involved in the defense of their country should be borne only by those who are taking their moral responsibilities seriously, and no reason why those who are not should benefit gratuitously. In the absence of argument showing that the method of individual initiative yields a more efficient army, the other seems to me preferable and, in any case, obviously fairer. Hesitation to use the lawmaking force of the community is understandable, for such use may endanger individual freedom, but often this hesitation is supported on the grounds of the moral preferability of individual sacrifice and initiative. Such arguments seem to me unsound.

12

Modern moral philosophy

G. E. M. ANSCOMBE

I will begin by stating three theses which I present in this paper. The first is that it is not profitable for us at present to do moral philosophy; that should be laid aside at any rate until we have an adequate philosophy of psychology, in which we are conspicuously lacking. The second is that the concepts of obligation, and duty – *moral* obligation and *moral* duty, that is to say – and of what is *morally* right and wrong, and of the *moral* sense of 'ought', ought to be jettisoned if this is psychologically possible; because they are survivals, or derivatives from survivals, from an earlier conception of ethics which no longer generally survives, and are only harmful without it. My third thesis is that the differences between the well-known English writers on moral philosophy from Sidgwick to the present day are of little importance.

Anyone who has read Aristotle's *Ethics* and has also read modern moral philosophy must have been struck by the great contrasts between them. The concepts which are prominent among the moderns seem to be lacking, or at any rate buried or far in the background, in Aristotle. Most noticeably, the term 'moral' itself, which we have by direct inheritance from Aristotle, just doesn't seem to fit, in its modern sense, into an account of Aristotelian ethics. Aristotle distinguishes virtues as moral and intellectual. Have some of what he calls 'intellectual' virtues what

---

Reprinted from Philosophy, *1958, pp. 1–19, by permission of the author and* Philosophy. *Relevant to this paper are J. Bennett, 'Whatever the consequences',* Analysis, *1965–6, and J. Cargile, 'On consequentialism',* Analysis, *1968–9*

we should call a 'moral' aspect? It would seem so; the criterion is presumably that a failure in an 'intellectual' virtue – like that of having good judgement in calculating how to bring about something useful, say in municipal government – may be *blameworthy*. But – it may reasonably be asked – cannot *any* failure be made a matter of blame or reproach? Any derogatory criticism, say of the workmanship of a product or the design of a machine, can be called blame or reproach. So we want to put in the word 'morally' again: sometimes such a failure may be *morally* blameworthy, sometimes not. Now has Aristotle got this idea of *moral* blame, as opposed to any other? If he has, why isn't it more central? There are some mistakes, he says, which are causes, not of involuntariness in actions, but of scoundrelism, and for which a man is blamed. Does this mean that there is a *moral* obligation not to make certain intellectual mistakes? Why doesn't he discuss obligation in general, and this obligation in particular? If someone professes to be expounding Aristotle and talks in a modern fashion about 'moral' such-and-such, he must be very imperceptive if he does not constantly feel like someone whose jaws have somehow got out of alignment: the teeth don't come together in a proper bite.

We cannot, then, look to Aristotle for any elucidation of the modern way of talking about 'moral' goodness, obligation, etc. And all the best-known writers on ethics in modern times, from Butler to Mill, appear to me to have faults as thinkers on the subject which make it impossible to hope for any direct light on it from them. I will state these objections with the brevity which their character makes possible.

Butler exalts conscience, but appears ignorant that a man's conscience may tell him to do the vilest things.

Hume defines 'truth' in such a way as to exclude ethical judgements from it, and professes that he has proved that they are so excluded. He also implicitly defines 'passion' in such a way that aiming at anything is having a passion. His objection to passing from 'is' to 'ought' would apply equally to passing from 'is' to 'owes' or from 'is' to 'needs'.[1] (However, because of the historical situation, he has a point here, which I shall return to.)

[1] David Hume *A Treatise of Human Nature*, Book III, Part I, Section 1. *Eds.*

Kant introduces the idea of 'legislating for oneself', which is as absurd as if in these days, when majority votes command great respect, one were to call each reflective decision a man made a vote resulting in a majority, which as a matter of proportion is overwhelming, for it is always 1–0. The concept of legislation requires superior power in the legislator. His own rigoristic convictions on the subject of lying were so intense that it never occurred to him that a lie could be relevantly described as anything but just a lie (e.g. as 'a lie in such-and-such circumstances'). His rule about universalizable maxims is useless without stipulations as to what shall count as a relevant description of an action with a view to constructing a maxim about it.

Bentham and Mill do not notice the difficulty of the concept 'pleasure'. They are often said to have gone wrong through committing the 'naturalistic fallacy'; but this charge does not impress me, because I do not find accounts of it coherent. But the other point – about pleasure – seems to me a fatal objection from the very outset. The ancients found this concept pretty baffling. It reduced Aristotle to sheer babble about 'the bloom on the cheek of youth' because, for good reasons, he wanted to make it out both identical with and different from the pleasurable activity.[1] Generations of modern philosophers found this concept quite unperplexing, and it reappeared in the literature as a problematic one only a year or two ago when Ryle wrote about it. The reason is simple: since Locke, pleasure was taken to be some sort of internal impression. But it was superficial, if that was the right account of it, to make it the point of actions. One might adapt something Wittgenstein said about 'meaning' and say 'Pleasure cannot be an internal impression, for no internal impression could have the consequences of pleasure.'

Mill also, like Kant, fails to realize the necessity for stipulation as to relevant descriptions, if his theory is to have content. It did not occur to him that acts of murder and theft could be otherwise described. He holds that where a proposed action is of such a kind as to fall under some one principle established on grounds of utility, one must go by that; where it falls under none or several, the several suggesting contrary views of the action, the thing to

[1] Aristotle, *Nicomachean Ethics* 1174b31. *Eds.*

do is to calculate particular consequences. But pretty well any action can be so described as to make it fall under a variety of principles of utility (as I shall say for short) if it falls under any.

I will now return to Hume. The features of Hume's philosophy which I have mentioned, like many other features of it, would incline me to think that Hume was a mere – brilliant – sophist; and his procedures are certainly sophistical. But I am forced, not to reverse, but to add to, this judgement by a peculiarity of Hume's philosophizing: namely that although he reaches his conclusions – with which he is in love – by sophistical methods, his considerations constantly open up very deep and important problems. It is often the case that in the act of exhibiting the sophistry one finds oneself noticing matters which deserve a lot of exploring: the obvious stands in need of investigation as a result of the points that Hume pretends to have made. In this, he is unlike, say, Butler. It was already well known that conscience could dictate vile actions; for Butler to have written disregarding this does not open up any new topics for us. But with Hume it is otherwise: hence he is a very profound and great philosopher, in spite of his sophistry. For example:

Suppose that I say to my grocer 'Truth consists in *either* relations of ideas, as that 20*s*. = £1, *or* matters of fact, as that I ordered potatoes, you supplied them, and you sent me a bill. So it doesn't apply to such a proposition as that I *owe* you such-and-such a sum.'

Now if one makes this comparison, it comes to light that the relation of the facts mentioned to the description 'X owes Y so much money' is an interesting one, which I will call that of being 'brute relative to' that description. Further, the 'brute' facts mentioned here themselves have descriptions relatively to which *other* facts are 'brute' – as, e.g., *he had potatoes carted to my house* and *they were left there* are brute facts relative to 'he supplied me with potatoes'. And the fact *X owes Y money* is in turn 'brute' relative to other descriptions – e.g. 'X is solvent'. Now the relation of 'relative bruteness' is a complicated one. To mention a few points: if xyz is a set of facts brute relative to a description A, then xyz is a set out of a range some set among which holds if A holds; but the holding of some set among these does not necessarily entail A, because exceptional circumstances can always make a difference; and what are exceptional circumstances relatively to A can generally only be explained by giving a few diverse examples, and *no* theoretically adequate provision can be made for exceptional circumstances, since a further special context can theoretically always be imagined that would reinterpret any special context. Further, though in normal circumstances, xyz would be a justification for A, that is not to say that A just comes to the same as 'xyz'; and also there is apt to be an institutional context which gives its point to the description A, of which institution A is of course not itself a description. (E.g. the statement that I give someone a shilling is not a description of the institution of money or of the currency of this country.) Thus, though it would be ludicrous to pretend that there can be no such thing as a transition from, e.g., 'is' to 'owes', the character of the transition is in fact rather interesting and comes to light as a result of reflecting on Hume's arguments.[1]

That I owe the grocer such-and-such a sum would be one of a set of facts which would be 'brute' in relation to the description 'I am a bilker'. 'Bilking' is of course a species of 'dishonesty' or 'injustice'. (Naturally the consideration will not have any effect on my actions unless I want to commit or avoid acts of injustice.)

So far, in spite of their strong association, I conceive 'bilking', 'injustice' and 'dishonesty' in a merely 'factual' way. That I can do this for 'bilking' is obvious enough; 'justice' I have no idea how to define, except that its sphere is that of actions which relate to someone else, but 'injustice', its defect, can for the moment be offered as a generic name covering various species. E.g.: 'bilking', 'theft' (which is relative to whatever property institutions exist), 'slander', 'adultery', 'punishment of the innocent'.

In present-day philosophy an explanation is required how an unjust man is a bad man, or an unjust action a bad one; to give

[1] The above two paragraphs are an abstract of a paper 'On Brute Facts', which appeared in *Analysis*, 1957–8. The paper is discussed in D. Z. Phillips, 'Miss Anscombe's grocer', *Analysis*, 1967–8, and C. Williamson, 'The grocers of Miss Anscombe and Mr Phillips', *Analysis*, 1967–8. *Eds.*

such an explanation belongs to ethics; but it cannot even be begun until we are equipped with a sound philosophy of psychology. For the proof that an unjust man is a bad man would require a positive account of justice as a 'virtue'. This part of the subject-matter of ethics is, however, completely closed to us until we have an account of what *type of characteristic* a virtue is – a problem, not of ethics, but of conceptual analysis – and how it relates to the actions in which it is instanced: a matter which I think Aristotle did not succeed in really making clear. For this we certainly need an account at least of what a human action is at all, and how its description as 'doing such-and-such' is affected by its motive and by the intention or intentions in it; and for this an account of such concepts is required.

The terms 'should' or 'ought' or 'needs' relate to good and bad: e.g. machinery needs oil, or should or ought to be oiled, in that running without oil is bad for it, or it runs badly without oil. According to this conception, of course, 'should' and 'ought' are not used in a special 'moral' sense when one says that a man should not bilk. (In Aristotle's sense of the term 'moral' (ἠθικός), they are being used in connection with a *moral* subject-matter: namely that of human passions and (non-technical) actions). But they have now acquired a special so-called 'moral' sense – i.e. a sense in which they imply some absolute verdict (like one of guilty/not guilty on a man) on what is described in the 'ought' sentences used in certain types of context: not merely the contexts that *Aristotle* would call 'moral' – passions and actions – but also some of the contexts that he would call 'intellectual'.

The ordinary (and quite indispensable) terms 'should', 'needs', 'ought', 'must' – acquired this special sense by being equated in the relevant contexts with 'is obliged', or 'is bound', or 'is required to', in the sense in which one can be obliged or bound by law, or something can be required by law.

How did this come about? The answer is in history: between Aristotle and us came Christianity, with its *law* conception of ethics. For Christianity derived its ethical notions from the Torah. (One might be inclined to think that a law conception of ethics could arise only among people who accepted an allegedly divine positive law; that this is not so is shown by the example of the Stoics, who also thought that whatever was involved in conformity to human virtues was required by divine law.)

In consequence of the dominance of Christianity for many centuries, the concepts of being bound, permitted, or excused became deeply embedded in our language and thought. The Greek word 'ἁμαρτάνειν', the aptest to be turned to that use, acquired the sense 'sin', from having meant 'mistake', 'missing the mark', 'going wrong'. The Latin *peccatum* which roughly corresponded to ἁμάρτημα was even apter for the sense 'sin', because it was already associated with '*culpa*' – 'guilt' – a juridical notion. The blanket term 'illicit', 'unlawful', meaning much the same as our blanket term 'wrong', explains itself. It is interesting that Aristotle did not have such a blanket term. He has blanket terms for wickedness – 'villain', 'scoundrel'; but of course a man is not a villain or a scoundrel by the performance of one bad action, or a few bad actions. And he has terms like 'disgraceful', 'impious'; and specific terms signifying defect of the relevant virtue, like 'unjust'; but no term corresponding to 'illicit'. The extension of this term (i.e. the range of its application) could be indicated in his terminology only by a quite lengthy sentence: that is 'illicit' which, whether it is a thought or a consented-to passion or an action or an omission in thought or action, is something contrary to one of the virtues the lack of which shows a man to be bad *qua* man. That formulation would yield a concept co-extensive with the concept 'illicit'.

To have a *law* conception of ethics is to hold that what is needed for conformity with the virtues failure in which is the mark of being bad *qua* man (and not merely, say, *qua* craftsman or logician) – that what is needed for *this*, is required by divine law. Naturally it is not possible to have such a conception unless you believe in God as a law-giver; like Jews, Stoics, and Christians. But if such a conception is dominant for many centuries, and then is given up, it is a natural result that the concepts of 'obligation', of being bound or required as by a law, should remain though they had lost their root; and if the word 'ought' has become invested in certain contexts with the sense of 'obligation', it too will remain to be spoken with a special emphasis and a special feeling in these contexts.

It is as if the notion 'criminal' were to remain when criminal law and criminal courts had been abolished and forgotten. A Hume discovering this situation might conclude that there was a special sentiment, expressed by 'criminal', which alone gave the word its sense. So Hume discovered the situation in which the notion 'obligation' survived, and the notion 'ought' was invested with that peculiar force having which it is said to be used in a 'moral' sense, but in which the belief in divine law had long since been abandoned: for it was substantially given up among Protestants at the time of the Reformation.[1] The situation, if I am right, was the interesting one of the survival of a concept outside the framework of thought that made it a really intelligible one.

When Hume produced his famous remarks about the transition from 'is' to 'ought', he was, then, bringing together several quite different points. One I have tried to bring out by my remarks on the transition from 'is' to 'owes' and on the relative 'bruteness' of facts. It would be possible to bring out a different point by enquiring about the transition from 'is' to 'needs'; from the characteristics of an organism to the environment that it needs, for example. To say that it needs that environment is not to say, e.g. that you want it to have that environment, but that it won't flourish unless it has it. Certainly, it all depends whether you *want* it to flourish! as Hume would say. But what 'all depends' on whether you want it to flourish is whether the fact that it needs that environment, or won't flourish without it, has the slightest influence on your actions. Now *that* such-and-such 'ought' to be or 'is needed' is supposed to have an influence on your actions: from which it seemed natural to infer that to judge that it 'ought to be' was in fact to grant what you judged 'ought to be' influence on your actions. And no amount of truth as to what *is* the case could possibly have a logical claim to have influence on your actions. (It is not judgement as such that sets us in motion; but

[1] They did not deny the existence of divine law; but their most characteristic doctrine was that it was given, not to be obeyed, but to show man's incapacity to obey it, even by grace; and this applied not merely to the ramified prescriptions of the Torah, but to the requirements of 'natural divine law'. Cf. in this connection the decree of Trent against the teaching that Christ was only to be trusted in as mediator, not obeyed as legislator.

our judgement on how to get or do something we *want.*) Hence it *must* be impossible to infer 'needs' or 'ought to be' from 'is'. But in the case of a plant, let us say, the inference from 'is' to 'needs' is certainly not in the least dubious. It is interesting and worth examining; but not at all fishy. Its interest is similar to the interest of the relation between brute and less brute facts: these relations have been very little considered. And while you can contrast 'what it needs' with 'what it's got' – like contrasting *de facto* and *de iure* – that does not make its needing this environment less of a 'truth'.

Certainly in the case of what the plant needs, the thought of a need will only affect action if you want the plant to flourish. Here, then, there is no necessary connection between what you can judge the plant 'needs' and what you want. But there is some sort of necessary connection between what you think *you* need, and what you want. The connection is a complicated one; it is possible *not* to want something that you judge you need. But, e.g., it is not possible never to want *anything* that you judge you need. This, however, is not a fact about the meaning of the word 'to need', but about the phenomenon of *wanting*. Hume's reasoning, we might say, in effect, leads one to think it must be about the word 'to need', or 'to be good for'.

Thus we find two problems already wrapped up in the remark about a transition from 'is' to 'ought'; now supposing that we had clarified the 'relative bruteness' of facts on the one hand, and the notions involved in 'needing', and 'flourishing' on the other – there would *still* remain a third point. For, following Hume, someone might say: Perhaps you have made out your point about a transition from 'is' to 'owes' and from 'is' to 'needs': but only at the cost of showing 'owes' and 'needs' sentences to express a *kind* of truths, a *kind* of facts. And it remains impossible to infer '*morally ought*' from 'is' sentences.

This comment, it seems to me, would be correct. This word 'ought', having become a word of mere mesmeric force, could not, in the character of having that force, be inferred from anything whatever. It may be objected that it could be inferred from other 'morally ought' sentences: but that cannot be true. The appearance that this is so is produced by the fact that we say 'All men are ϕ' and 'Socrates is a man' implies 'Socrates is ϕ'. But here 'ϕ' is a dummy predicate. We mean that if you substitute a real predicate for 'ϕ' the implication is valid. A real predicate is required; not just a word containing no intelligible thought: a word retaining the suggestion of force, and apt to have a strong psychological effect, but which no longer signifies a real concept at all.

For its suggestion is one of a *verdict* on my action, according as it agrees or disagrees with the description in the 'ought' sentence. And where one does not think there is a judge or a law, the notion of a verdict may retain its psychological effect, but not its meaning. Now imagine that just this word 'verdict' *were* so used – with a characteristically solemn emphasis – as to retain its atmosphere but not its meaning, and someone were to say: 'For a *verdict*, after all, you need a law and a judge.' The reply might be made: 'Not at all, for if there were a law and a judge who gave a verdict, the question for us would be whether accepting that verdict is something that there is a *Verdict* on.' This is an analogue of an argument which is so frequently referred to as decisive: If someone does have a divine law conception of ethics, all the same, he has to agree that he has to have a judgement that he *ought* (morally ought) to obey the divine law; so his ethic is in exactly the same position as any other: he merely has a 'practical major premise'[1]: 'Divine law ought to be obeyed' where someone else has, e.g., 'The greatest happiness principle ought to be employed in all decisions.'

I should judge that Hume and our present-day ethicists had done a considerable service by showing that no content could be found in the notion 'morally ought'; if it were not that the latter philosophers try to find an alternative (very fishy) content and to retain the psychological force of the term. It would be most reasonable to drop it. It has no reasonable sense outside a law conception of ethics; they are not going to maintain such a conception; and you can do ethics without it, as is shown by the example of Aristotle. It would be a great improvement, if, instead of

[1] As it is absurdly called. Since major premise = premise containing the term which is predicate in the conclusion, it is a solecism to speak of it in the connection with practical reasoning.

'morally wrong', one always named a genus such as 'untruthful', 'unchaste', 'unjust'. We should no longer ask whether doing something was 'wrong', passing directly from some description of an action to this notion; we should ask whether, e.g., it was unjust; and the answer would sometimes be clear at once.

I now come to the epoch in modern English moral philosophy marked by Sidgwick. There is a startling change that seems to have taken place between Mill and Moore. Mill assumes, as we saw, that there is no question of calculating the particular consequences of an action such as murder or theft; and we saw too that his position was stupid, because it is not at all clear how an action *can* fall under just one principle of utility. In Moore and in subsequent academic moralists of England we find it taken to be pretty obvious that 'the right action' is the action which produces the best possible consequences (reckoning among consequences the intrinsic values ascribed to certain kinds of act by some 'Objectivists'[1]). Now it follows from this that a man does well, subjectively speaking, if he acts for the best in the particular circumstances according to his judgement of the total consequences of this particular action. I say that this follows, not that any philosopher has said precisely that. For discussion of these questions can of course get extremely complicated: e.g. it can be doubted whether 'such-and-such is the right action' is a satisfactory formulation, on the grounds that things have to exist to have predicates – so perhaps the best formulation is 'I am obliged'; or again, a philosopher may deny that 'right' is a 'descriptive' term, and then take a roundabout route through linguistic analysis to reach a view which comes to the same thing as 'the right action is the one productive of the best consequences' (e.g. the view that you frame your 'principles' to effect the end you choose to pursue, the connexion between 'choice' and 'best' being supposedly such that choosing reflectively means that you choose how to act so as to produce the best consequences); further, the roles of what are called 'moral principles' and of the

[1] Oxford Objectivists of course distinguish between 'consequences' and 'intrinsic values' and so produce a misleading appearance of not being 'consequentialists'. But they do not hold – and Ross explicitly denies – that the gravity of, e.g., procuring the condemnation of the innocent is such that it cannot be outweighed by, e.g., national interest. Hence their distinction is of no importance.

'motive of duty' have to be described; the differences between 'good' and 'morally good' and 'right' need to be explored, the special characteristics of 'ought' sentences investigated. Such discussions generate an appearance of significant diversity of views where what is really significant is an overall similarity. The overall similarity is made clear if you consider that every one of the best known English academic moral philosophers has put out a philosophy according to which, e.g., it is not possible to hold that it cannot be right to kill the innocent as a means to any end whatsoever and that someone who thinks otherwise is in error. (I have to mention both points; because Mr Hare, for example, while teaching a philosophy which would encourage a person to judge that killing the innocent would be what he 'ought' to choose for over-riding purposes, would also teach, I think, that if a man chooses to make avoiding killing the innocent for any purpose his 'supreme practical principle', he cannot be impugned for error: that just is his 'principle'. But with that qualification, I think it can be seen that the point I have mentioned holds good of every single English academic moral philosopher since Sidgwick.) Now this is a significant thing: for it means that all these philosophies are quite incompatible with the Hebrew-Christian ethic. For it has been characteristic of that ethic to teach that there are certain things forbidden whatever *consequences* threaten, such as: choosing to kill the innocent for any purpose, however good; vicarious punishment; treachery (by which I mean obtaining a man's confidence in a grave matter by promises of trustworthy friendship and then betraying him to his enemies); idolatry; sodomy; adultery; making a false profession of faith. The prohibition of certain things simply in virtue of their description as such-and-such identifiable kinds of action, regardless of any further consequences, is certainly not the whole of the Hebrew-Christian ethic; but it is a noteworthy feature of it; and if every academic philosopher since Sidgwick has written in such a way as to exclude this ethic, it would argue a certain provinciality of mind not to see this incompatibility as the most important fact about these philosophers, and the differences between them as somewhat trifling by comparison.

It is noticeable that none of these philosophers displays any consciousness that there is such an ethic, which he is contradicting: it is pretty well taken for obvious among them all that a prohibition such as that on murder does not operate in face of some consequences. But of course the strictness of the prohibition has as its point *that you are not to be tempted by fear or hope of consequences.*

If you notice the transition from Mill to Moore, you will suspect that it was made somewhere by someone; Sidgwick will come to mind as a likely name; and you will in fact find it going on, almost casually, in him. He is rather a dull author; and the important things in him occur in asides and footnotes and small bits of argument which are not concerned with his grand classification of the 'methods of ethics'. A divine law theory of ethics is reduced to an insignificant variety by a footnote telling us that 'the best theologians' (God knows whom he meant) tell us that God is to be obeyed in his capacity of a *moral* being. ἦ φορτικὸς ὁ ἔπαινος; one seems to hear Aristotle saying: 'Isn't the praise vulgar?'[1] – But Sidgwick *is* vulgar in that kind of way: he thinks, for example, that humility consists in underestimating your own merits – i.e. in a species of untruthfulness; and that the ground for having laws against blasphemy was that it was offensive to believers; and that to go accurately into the virtue of purity is to offend against its canons, a thing he reproves 'medieval theologians' for not realizing.

From the point of view of the present enquiry, the most important thing about Sidgwick was his definition of intention. He defines intention in such a way that one must be said to intend any foreseen consequences of one's voluntary action. This definition is obviously incorrect, and I dare say that no one would be found to defend it now. He uses it to put forward an ethical thesis which would now be accepted by many people: the thesis that it does not make any difference to a man's responsibility for something that he foresaw, that he felt no desire for it, either as an end or as a means to an end. Using the language of intention more correctly, and avoiding Sidgwick's faulty conception, we may state the thesis thus: it does not make any difference to a

[1] E.N. 1178b16.

man's responsibility for an effect of his action which he can foresee, that he does not intend it. Now this sounds rather edifying; it is I think quite characteristic of very bad degenerations of thought on such questions that they sound edifying. We can see what it amounts to by considering an example. Let us suppose that a man has a responsibility for the maintenance of some child. Therefore deliberately to withdraw support from it is a bad sort of thing for him to do. It would be bad for him to withdraw its maintenance because he didn't want to maintain it any longer; *and* also bad for him to withdraw it because by doing so he would, let us say, compel someone else to do something. (We may suppose for the sake of argument that compelling that person to do that thing is in itself quite admirable.) But now he has to choose between doing something disgraceful and going to prison; if he goes to prison, it will follow that he withdraws support from the child. By Sidgwick's doctrine, there is no difference in his responsibility for ceasing to maintain the child, between the case where he does it for its own sake or as a means to some other purpose, and when it happens as a foreseen and unavoidable consequence of his going to prison rather than do something disgraceful. It follows that he must weigh up the relative badness of withdrawing support from the child and of doing the disgraceful thing; and it may easily be that the disgraceful thing is in fact a less vicious action than intentionally withdrawing support from the child would be; if then the fact that withdrawing support from the child is a side effect of his going to prison does not make any difference to his responsibility, this consideration will incline him to do the disgraceful thing; which can still be pretty bad. And of course, once he has started to look at the matter in this light, the only reasonable thing for him to consider will be the consequences and not the intrinsic badness of this or that action. So that, given that he judges reasonably that no *great* harm will come of it, he can do a much more disgraceful thing than deliberately withdrawing support from the child. And if his calculations turn out in fact wrong, it will appear that he was not responsible for the consequences, because he did not foresee them. For in fact Sidgwick's thesis leads to its being quite impossible to estimate the badness of an action except in the light of *expected* consequences. But if so, then *you* must estimate the badness in the light of the consequences *you* expect; and so it will follow that you can exculpate yourself from the *actual* consequences of the most disgraceful actions, so long as you can make out a case for not having foreseen them. Whereas I should contend that a man is responsible for the bad consequences of his bad actions, but gets no credit for the good ones; and contrariwise is not responsible for the bad consequences of good actions.

The denial of *any* distinction between foreseen and intended consequences, as far as responsibility is concerned, was not made by Sidgwick in developing any one 'method of ethics'; he made this important move on behalf of everybody and just on its own account; and I think it plausible to suggest that *this* move on the part of Sidgwick explains the difference between old-fashioned Utilitarianism and that *consequentialism*, as I name it, which marks him and every English academic moral philosopher since him. By it, the kind of consideration which would formerly have been regarded as a temptation, the kind of consideration urged upon men by wives and flattering friends, was given a status by moral philosophers in their theories.

It is a necessary feature of consequentialism that it is a shallow philosophy. For there are always borderline cases in ethics. Now if you are either an Aristotelian, or a believer in divine law, you will deal with a borderline case by considering whether doing such-and-such in such-and-such circumstances is, say, murder, or is an act of injustice; and according as you decide it is or it isn't, you judge it to be a thing to do or not. This would be the method of casuistry; and while it may lead you to stretch a point on the circumference, it will not permit you to destroy the centre. But if you are a consequentialist, the question 'What is it right to do in such-and-such circumstances?' is a stupid one to raise. The casuist raises such a question only to ask 'Would it be *permissible* to do so-and-so?' or 'Would it be permissible *not* to do so-and-so?' Only if it would *not* be permissible *not* to do so-and-so could he say '*This* would be *the* thing to do'.[1] Otherwise, though he

[1] Necessarily a rare case: for the positive precepts, e.g. 'Honour your parents', hardly ever prescribe, and seldom even necessitate, any particular action.

may speak *against* some action, he cannot prescribe any – for in an *actual* case, the circumstances (beyond the ones imagined) might suggest all sorts of possibilities, and you can't know in advance what the possibilities are going to be. Now the consequentialist has no footing on which to say 'This would be permissible, this not'; because by his own hypothesis, it is the consequences that are to decide, and he has no business to pretend that he can lay it down what possible twists a man could give doing this or that; the most he can say is: a man must not *bring about* this or that; he has no right to say he will, in an actual case, bring about such-and-such unless he does so-and-so. Further, the consequentialist, in order to be imagining borderline cases at all, has of course to assume some sort of law or standard according to which this is a borderline case. Where then does he get the standard from? In practice the answer invariably is: from the standards current in his society or his circle. And it has in fact been the mark of all these philosophers that they have been extremely conventional; they have nothing in them by which to revolt against the conventional standards of their sort of people; it is impossible that they should be profound. But the chance that a whole range of conventional standards will be decent is small. Finally, the point of considering hypothetical situations, perhaps very improbable ones, *seems* to be to elicit from yourself or someone else a hypothetical decision to do something of a bad kind. I don't doubt this has the effect of predisposing people – who will never get into the situations for which they have made hypothetical choices – to consent to similar bad actions, or to praise and flatter those who do them, so long as their crowd does so too, when the desperate circumstances imagined don't hold at all.

Those who recognize the origins of the notions of 'obligation' and of the emphatic, 'moral', *ought*, in the divine law conception of ethics, but who reject the notion of a divine legislator, sometimes look about for the possibility of retaining a law conception without a divine legislator. This search, I think, has some interest in it. Perhaps the first thing that suggests itself is the 'norms' of a society. But just as one cannot be impressed by Butler when one reflects what conscience can tell people to do, so, I think, one cannot be impressed by this idea if one reflects what the 'norms' of a society can be like. That legislation can be 'for oneself' I reject as absurd; whatever you do 'for yourself' may be admirable; but is not legislating. Once one sees this, one may say: I have to frame my own rules, and these are the best I can frame, and I shall go by them until I know something better: as a man might say 'I shall go by the customs of my ancestors'. Whether this leads to good or evil will depend on the *content* of the rules or of the customs of one's ancestors. If one is lucky it will lead to good. Such an attitude would be hopeful in this at any rate: it seems to have in it some Socratic doubt where, from having to fall back on such expedients, it should be clear that Socratic doubt is good; in fact rather generally it must be good for anyone to think 'Perhaps in some way I can't see, I may be on a bad path, perhaps I am hopelessly wrong in some essential way'. The search for 'norms' might lead someone to look for laws of nature, as if the universe were a legislator; but in the present day this is not likely to lead to good results: it might lead one to eat the weaker according to the laws of nature, but would hardly lead anyone nowadays to notions of justice; the pre-Socratic feeling about justice as comparable to the balance or harmony which kept things going is very remote to us.

There is another possibility here: 'obligation' may be contractual. Just as we look at the law to find out what a man subject to it is required by it to do, so we look at a contract to find out what the man who has made it is required by it to do. Thinkers, admittedly remote from us, might have the idea of a *foedus rerum*, of the universe not as a legislator but as the embodiment of a contract. Then if you could find out what the contract was, you would learn your obligations under it. Now, you cannot be under a law unless it has been promulgated to you; and the thinkers who believed in 'natural divine law' held that it was promulgated to every grown man in his knowledge of good and evil. Similarly you cannot be in a contract without having contracted, i.e. given signs of entering upon the contract. Just possibly, it might be argued that the use of language which one makes in the ordinary conduct of life amounts in some sense to giving the signs of entering into various contracts. If anyone had this theory, we should want to see it worked out. I suspect that it would be largely formal; it might be possible to construct a system embodying the law (whose status might be compared to that of 'laws' of logic): 'what's sauce for the goose is sauce for the gander', but hardly one descending to such particularities as the prohibition on murder or sodomy. Also, while it is clear that you can be subject to a law that you do not acknowledge and have not thought of as law, it does not seem reasonable to say that you can enter upon a contract without knowing that you are doing so; such ignorance is usually held to be destructive of the nature of a contract.

It might remain to look for 'norms' in human virtues: just as *man* has so many teeth, which is certainly not the average number of teeth men have, but is the number of teeth for the species, so perhaps the species *man*, regarded not just biologically, but from the point of view of the activity of thought and choice in regard to the various departments of life – powers and faculties and use of things needed – 'has' such-and-such virtues: and this 'man' with the complete set of virtues is the 'norm', as 'man' with, e.g., a complete set of teeth is a norm. But in *this* sense 'norm' has ceased to be roughly equivalent to 'law'. In *this* sense the notion of a 'norm' brings us nearer to an Aristotelian than a law conception of ethics. There is, I think, no harm in that; but if someone looked in this direction to give 'norm' a sense, then he ought to recognize what has happened to the notion 'norm', which he wanted to mean 'law – without bringing God in' – it has ceased to mean 'law' at all; and *so* the notions of 'moral obligation', 'the moral ought', and 'duty' are best put on the Index, if he can manage it.

But meanwhile – is it not clear that there are several concepts that need investigating simply as part of the philosophy of psychology and – as I should recommend – *banishing ethics totally* from our minds? Namely – to begin with: 'action', 'intention', 'pleasure', 'wanting'. More will probably turn up if we start with these. Eventually it might be possible to advance to considering the concept 'virtue'; with which, I suppose, we should be beginning some sort of a study of ethics.

I will end by describing the advantages of using the word 'ought' in a non-emphatic fashion, and not in a special 'moral' sense; of discarding the term 'wrong' in a 'moral' sense, and using such notions as 'unjust'.

It is possible, if one is allowed to proceed just by giving examples, to distinguish between the intrinsically unjust, and what is unjust given the circumstances. To arrange to get a man judicially punished for something which it can be clearly seen he has not done is intrinsically unjust. This might be done, of course, and often has been done, in all sorts of ways; by suborning false witnesses, by a rule of law by which something is 'deemed' to be the case which is admittedly not the case as a matter of fact, and by open insolence on the part of the judges and powerful people when they more or less openly say: 'A fig for the fact that you did not do it; we mean to sentence you for it all the same.' What is unjust given, e.g., normal circumstances is to deprive people of their ostensible property without legal procedure, not to pay debts, not to keep contracts, and a host of other things of the kind. Now, the circumstances can clearly make a great deal of difference in estimating the justice or injustice of such procedures as these; and these circumstances may *sometimes* include expected consequences; for example, a man's claim to a bit of property can become a nullity when its seizure and use can avert some obvious disaster: as, e.g., if you could use a machine of his to produce an explosion in which it would be destroyed, but by means of which you could divert a flood or make a gap which a fire could not jump. Now this certainly does not mean that what would ordinarily be an act of injustice, but is not intrinsically unjust, can always be rendered just by a reasonable calculation of better consequences; far from it; but the problems that would be raised in an attempt to draw a boundary line (or boundary area) here are obviously complicated. And while there are certainly some general remarks which ought to be made here, and some boundaries that can be drawn, the decision on particular cases would for the most part be determined *κατὰ τὸν ὀρθὸν λόγον*, 'according to what's reasonable'. E.g. that *such-and-such* a delay of payment of a *such-and-such* debt to a person *so* circumstanced, on the part of a person *so* circumstanced, would or would not be unjust, is really only to be decided 'according to what's reasonable'; and for this there can *in principle* be no canon other than giving a few examples. That is to say, while it is because of a big gap in philosophy that we can give no general account of the concept of virtue and of the concept of justice, but have to proceed, using the concepts, only by giving examples; still there is an area where it is not because of any gap, but is in principle the case, that there is no account except by way of examples: and that is where the canon is 'what's reasonable': which of course is *not* a canon.

That is all I wish to say about what is just in some circumstances, unjust in others; and about the way in which expected consequences can play a part in determining what is just. Returning to my example of the intrinsically unjust: if a procedure *is* one of judicially punishing a man for what he is clearly understood not to have done, there can be absolutely no argument about the description of this as unjust. No circumstances, and no expected consequences, which do *not* modify the description of the procedure as one of judicially punishing a man for what he is known not to have done can modify the description of it as unjust. Someone who attempted to dispute this would only be pretending not to know what 'unjust' means: for this is a paradigm case of injustice.

And here we see the superiority of the term 'unjust' over the terms 'morally right' and 'morally wrong'. For in the context of English moral philosophy since Sidgwick it appears legitimate to discuss whether it *might* be 'morally right' in some circumstances to adopt that procedure; but it cannot be argued that the procedure would in any circumstances be just.

Now I am not able to do the philosophy involved – and I think that no one in the present situation of English philosophy *can* do the philosophy involved – but it is clear that a good man is a just man; and a just man is a man who habitually refuses to commit or participate in any unjust actions for fear of any consequences, or to obtain any advantage, for himself or anyone else. Perhaps no one will disagree. But, it will be said, what *is* unjust is sometimes determined by expected consequences; and certainly that is true. But there are cases where it is not: now if someone says, 'I agree, but all this wants a lot of explaining', then he is right, and, what is more, the situation at present is that we can't do the explaining; we lack the philosophic equipment. But if someone really thinks, *in advance,*[1] that it is open to question whether such an action as procuring the judicial execution of the innocent should be quite excluded from consideration – I do not want to argue with him; he shows a corrupt mind.

In such cases our moral philosophers seem to impose a dilemma upon us. 'If we have a case where the term "unjust" applies purely in virtue of a factual description, can't one raise the question whether one sometimes conceivably ought to do injustice? If "what is unjust" is determined by consideration of whether it is *right* to do so-and-so in such-and-such circumstances, then the question whether it is "right" to commit injustice can't arise, just because "wrong" has been built into the definition of injustice. But if we have a case where the description "unjust" applies purely in virtue of the facts, without bringing "wrong" in, then the question can arise whether one "ought" perhaps to commit an injustice, whether it might not be "right" to? And of course "ought" and "right" are being used in their *moral* senses here. Now either you must decide what is "morally right" in the light of certain *other* "principles", or you make a "principle" about *this* and decide that an injustice is never "right"; but even if you do the latter you are going beyond the facts; you are making a decision that you will not, or that it is wrong to, commit injustice. But in either case, *if* the term "unjust" is determined simply by the facts, it is not the term "unjust" that determines that the term "wrong" applies, but a decision that injustice is *wrong*, together with the diagnosis of the "factual" description as entailing injustice. But the man who makes an absolute decision that injustice is "wrong" has no footing on which to criticize someone who does *not* make that decision as judging falsely.'

[1] If he thinks it in the concrete situation, he is of course merely a normally tempted human being. In discussion when this paper was read, as was perhaps to be expected, this case was produced: a government is required to have an innocent man tried, sentenced and executed under threat of a 'hydrogen bomb war'. It would seem strange to me to have much hope of so averting a war threatened by such men as made this demand. But the most important thing about the way in which cases like this are invented in discussions, is the assumption that only two courses are open: here, compliance and open defiance. No one can say in advance of such a situation what the possibilities are going to be – e.g. that there is none of stalling by a feigned willingness to comply, accompanied by a skilfully arranged 'escape' of the victim.

In this argument 'wrong' of course is explained as meaning 'morally wrong', and all the atmosphere of the term is retained while its substance is guaranteed quite null. Now let us remember that 'morally wrong' is the term which is the heir of the notion 'illicit', or 'what there is an obligation *not* to do'; which belongs in a divine law theory of ethics. Here it really does add something to the description 'unjust' to say there is an obligation not to do it; for what obliges is the divine law – as rules oblige in a game. So if the divine law obliges not to commit injustice by forbidding injustice, it really does add something to the description 'unjust' to say there is an obligation not to do it. And it is because 'morally wrong' is the heir of this concept, but an heir that is cut off from the family of concepts from which it sprang, that 'morally wrong' *both* goes beyond the mere factual description 'unjust' *and* seems to have no discernible content except a certain compelling force, which I should call purely psychological. And such is the force of the term that philosophers actually suppose that the divine law notion can be dismissed as making no essential difference, even if it is held – *because* they think that a 'practical principle' running 'I *ought* (i.e. am morally obliged) to obey divine laws' is required for the man who believes in divine laws. But actually this notion of obligation is a notion which only operates in the context of law. And I should be inclined to congratulate the present-day moral philosophers on depriving 'morally ought' of its now delusive appearance of content, if only they did not manifest a detestable desire to retain the atmosphere of the term.

It may be possible, if we are resolute, to discard the notion 'morally ought', and simply return to the ordinary 'ought', which, we ought to notice, is such an extremely frequent term of human language that it is difficult to imagine getting on without it. Now if we do return to it, can't it reasonably be asked whether one might ever need to commit injustice, or whether it won't be the best thing to do? Of course it can. And the answers will be various. One man – a philosopher – may say that since justice is a virtue, and injustice a vice, and virtues and vices are built up by the performances of the action in which they are instanced, an act of injustice will tend to make a man bad; and essentially the flourishing of a man *qua* man consists in his being good (e.g. in virtues); but for any X to which such terms apply, X needs what makes it flourish, so a man needs, or ought to perform, only virtuous actions; and even if, as it must be admitted may happen, he flourishes less, or not at all, in inessentials, by avoiding injustice, his life is spoiled in essentials by not avoiding injustice – so he still needs to perform only just actions. That is roughly how Plato and Aristotle talk; but it can be seen that philosophically there is a huge gap, at present unfillable as far as we are concerned, which needs to be filled by an account of human nature, human action, the type of characteristic a virtue is, and above all of human 'flourishing'. And it is the last concept that appears the most doubtful. For it is a bit much to swallow that a man in pain and hunger and poor and friendless is 'flourishing', as Aristotle himself admitted. Further, someone might say that one at least needed to stay alive to 'flourish'. Another man unimpressed by all that will say in a hard case 'What we need is such-and-such, which we won't get without doing this (which is unjust) – so this is what we ought to do.' Another man, who does not follow the rather elaborate reasoning of the philosophers, simply says 'I know it is in any case a disgraceful thing to say that one had better commit this unjust action.' The man who believes in divine laws will say perhaps 'It is forbidden, and however it looks, it cannot be to anyone's profit to commit injustice'; he like the Greek philosophers can think in terms of 'flourishing'. If he is a Stoic, he is apt to have a decidedly strained notion of what 'flourishing' consists in; if he is a Jew or Christian, he need not have any very distinct notion: the way it will profit him to abstain from injustice is something that he leaves it to God to determine, himself only saying 'It can't do me any good to go against his law'. (But he also hopes for a great reward in a new life later on, e.g. at the coming of Messiah; but in this he is relying on special promises.)

It is left to modern moral philosophy – the moral philosophy of all the well-known English ethicists since Sidgwick – to construct systems according to which the man who says 'We need such-and-such, and will only get it this way' *may* be a virtuous character: that is to say, it is left open to debate whether such a procedure as the judicial punishment of the innocent may not in some circumstances be the 'right' one to adopt; and though the present Oxford moral philosophers would accord a man *permission* to 'make it his principle' not to do such a thing, they teach a philosophy according to which the particular consequences of such an action *could* 'morally' be taken into account by a man who was debating what to do; and if they were such as to conflict with his 'ends', it might be a step in his moral education to frame a moral principle under which he 'managed' (to use Mr Nowell-Smith's phrase[1]) to bring the action; or it might be a new 'decision of principle', making which was an advance in the formation of his moral thinking (to adopt Mr Hare's conception), to decide: in such-and-such circumstances one ought to procure the judicial condemnation of the innocent. And that is my complaint.

[1] *Ethics*, p. 308.

13

Morality and advantage

D. P. GAUTHIER

I

Hume asks, rhetorically, 'what theory of morals can ever serve any useful purpose, unless it can show, by a particular detail, that all the duties which it recommends, are also the true interest of each individual?'[1] But there are many to whom this question does not seem rhetorical. Why, they ask, do we speak the language of morality, impressing upon our fellows their duties and obligations, urging them with appeals to what is right and good, if we could speak to the same effect in the language of prudence, appealing to considerations of interest and advantage? When the poet, Ogden Nash, is moved by the muse to cry out:

> O Duty,
> Why hast thou not the visage of a sweetie or a cutie?[2]

we do not anticipate the reply:

> O Poet,
> I really am a cutie and I think you ought to know it.

The belief that duty cannot be reduced to interest, or that morality may require the agent to subordinate all considerations of advantage, is one which has withstood the assaults of contrary-minded philosophers from Plato to the present. Indeed, were it

[1] David Hume, *An Enquiry Concerning the Principles of Morals*, sec. ix, pt. ii.
[2] Ogden Nash, 'Kind of an Ode to Duty'.

Reprinted from Philosophical Review, *1967, pp. 460–75, by permission of the author and the* Philosophical Review.

not for the conviction that only interest and advantage can motivate human actions, it would be difficult to understand philosophers contending so vigorously for the identity, or at least compatibility, of morality with prudence.

Yet if morality is not true prudence it would be wrong to suppose that those philosophers who have sought some connection between morality and advantage have been merely misguided. For it is a truism that we should all expect to be worse off if men were to substitute prudence, even of the most enlightened kind, for morality in all of their deliberations. And this truism demands not only some connection between morality and advantage, but a seemingly paradoxical connection. For if we should all expect to suffer, were men to be prudent instead of moral, then morality must contribute to advantage in a unique way, a way in which prudence – following reasons of advantage – cannot.

Thomas Hobbes is perhaps the first philosopher who tried to develop this seemingly paradoxical connection between morality and advantage. But since he could not admit that a man might ever reasonably subordinate considerations of advantage to the dictates of obligation, he was led to deny the possibility of real conflict between morality and prudence. So his argument fails to clarify the distinction between the view that claims of obligation reduce to considerations of interest and the view that claims of obligation promote advantage in a way in which considerations of interest cannot.

More recently, Kurt Baier has argued that 'being moral is following rules designed to overrule self-interest whenever it is in the interest of everyone alike that everyone should set aside his interest'.[1] Since prudence is following rules of (enlightened) self-interest, Baier is arguing that morality is designed to overrule prudence when it is to everyone's advantage that it do so – or, in other words, that morality contributes to advantage in a way in which prudence cannot.[2]

[1] Kurt Baier, *The Moral Point of View: A Rational Basis of Ethics* (Ithaca, 1958), p. 314.

[2] That this, and only this, is what he is entitled to claim may not be clear to Baier, for he supposes his account of morality to answer the question 'Why should we be moral?', interpreting 'we' distributively. This, as I shall argue in Sec. IV, is quite mistaken.

Baier does not actually demonstrate that morality contributes to advantage in this unique and seemingly paradoxical way. Indeed, he does not ask how it is possible that morality should do this. It is this possibility which I propose to demonstrate.

II

Let us examine the following proposition, which will be referred to as 'the thesis': *Morality is a system of principles such that it is advantageous for everyone if everyone accepts and acts on it, yet acting on the system of principles requires that some persons perform disadvantageous acts.*[1]

What I wish to show is that this thesis *could be true*, that morality could possess those characteristics attributed to it by the thesis. I shall not try to show that the thesis is true – indeed, I shall argue in Section V that it presents at best an inadequate conception of morality. But it is plausible to suppose that a modified form of the thesis states a necessary, although not a sufficient, condition for a moral system.

Two phrases in the thesis require elucidation. The first is 'advantageous for everyone'. I use this phrase to mean that *each* person will do better if the system is accepted and acted on than if *either* no system is accepted and acted on *or* a system is accepted and acted on which is similar, save that it never requires any person to perform disadvantageous acts.

Clearly, then, the claim that it is advantageous for everyone to accept and act on the system is a very strong one; it may be so strong that no system of principles which might be generally adopted could meet it. But I shall consider in Section V one among the possible ways of weakening the claim.

The second phrase requiring elucidation is 'disadvantageous acts'. I use this phrase to refer to acts which, in the context of their performance, would be less advantageous to the performer than some other act open to him in the same context. The phrase

[1] The thesis is not intended to state Baier's view of morality. I shall suggest in Sec. V that Baier's view would require substituting 'everyone can expect to benefit' for 'it is advantageous to everyone'. The thesis is stronger and easier to discuss.

does not refer to acts which merely impose on the performer some short-term disadvantage that is recouped or outweighed in the long run. Rather it refers to acts which impose a disadvantage that is never recouped. It follows that the performer may say to himself, when confronted with the requirement to perform such an act, that it would be better *for him* not to perform it.

It is essential to note that the thesis, as elucidated, does not maintain that morality is advantageous for everyone in the sense that each person will do *best* if the system of principles is accepted and acted on. Each person will do better than if no system is adopted, or than if the one particular alternative mentioned above is adopted, but not than if any alternative is adopted.

Indeed, for each person required by the system to perform some disadvantageous act, it is easy to specify a better alternative – namely, the system modified so that it does not require *him* to perform any act disadvantageous to himself. Of course, there is no reason to expect such an alternative to be better than the moral system for everyone, or in fact for anyone other than the person granted the special exemption.

A second point to note is that each person must gain more from the disadvantageous acts performed by others than he loses from the disadvantageous acts performed by himself. If this were not the case, then some person would do better if a system were adopted exactly like the moral system save that it never requires *any* person to perform disadvantageous acts. This is ruled out by the force of 'advantageous for everyone'.

This point may be clarified by an example. Suppose that the system contains exactly one principle. Everyone is always to tell the truth. It follows from the thesis that each person gains more from those occasions on which others tell the truth, even though it is disadvantageous to them to do so, than he loses from those occasions on which he tells the truth even though it is disadvantageous to him to do so.

Now this is not to say that each person gains by telling others the truth in order to ensure that in return they tell him the truth. Such gains would merely be the result of accepting certain short-term disadvantages (those associated with truth-telling) in order to reap long-term benefits (those associated with being told the truth). Rather, what is required by the thesis is that those disadvantages which a person incurs in telling the truth, when he can expect neither short-term nor long-term benefits to accrue to him from truth-telling, are outweighed by those advantages he receives when others tell him the truth when they can expect no benefits to accrue to them from truth-telling.

The principle enjoins truth-telling in those cases in which whether one tells the truth or not will have no effect on whether others tell the truth. Such cases include those in which others have no way of knowing whether or not they are being told the truth. The thesis requires that the disadvantages one incurs in telling the truth in these cases are less than the advantages one receives in being told the truth by others in parallel cases; and the thesis requires that this holds for everyone.

Thus we see that although the disadvantages imposed by the system on any person are less than the advantages secured him through the imposition of disadvantages on others, yet the disadvantages are real in that incurring them is *unrelated* to receiving the advantages. The argument of long-term prudence, that I ought to incur some immediate disadvantage *so that* I shall receive compensating advantages later on, is entirely inapplicable here.

III

It will be useful to examine in some detail an example of a system which possesses those characteristics ascribed by the thesis to morality. This example, abstracted from the field of international relations, will enable us more clearly to distinguish, first, conduct based on immediate interest; second, conduct which is truly prudent; and third, conduct which promotes mutual advantage but is not prudent.

A and *B* are two nations with substantially opposed interests, who find themselves engaged in an arms race against each other. Both possess the latest in weaponry, so that each recognizes that the actual outbreak of full-scale war between them would be mutually disastrous. This recognition leads *A* and *B* to agree that each would be better off if they were mutually disarming instead of mutually arming. For mutual disarmament would preserve the balance of power between them while reducing the risk of war.

Hence *A* and *B* enter into a disarmament pact. The pact is advantageous for both if both accept and act on it, although clearly it is not advantageous for either to act on it if the other does not.

Let *A* be considering whether or not to adhere to the pact in some particular situation, whether or not actually to perform some act of disarmament. *A* will quite likely consider the act to have disadvantageous consequences. *A* expects to benefit, not by its own acts of disarmament, but by *B*'s acts. Hence if *A* were to reason simply in terms of immediate interest, *A* might well decide to violate the pact.

But *A*'s decision need be neither prudent nor reasonable. For suppose first that *B* is able to determine whether or not *A* adheres to the pact. If *A* violates, then *B* will detect the violation and will then consider what to do in the light of *A*'s behaviour. It is not to *B*'s advantage to disarm alone; *B* expects to gain, not by its own acts of disarmament, but by *A*'s acts. Hence *A*'s violation, if known to *B*, leads naturally to *B*'s counter-violation. If this continues, the effect of the pact is entirely undone, and *A* and *B* return to their mutually disadvantageous arms race. *A*, foreseeing this when considering whether or not to adhere to the pact in the given situation, must therefore conclude that the truly prudent course of action is to adhere.

Now suppose that *B* is unable to determine whether or not *A* adheres to the pact in the particular situation under consideration. If *A* judges adherence to be in itself disadvantageous, then it will decide, both on the basis of immediate interest and on the basis of prudence, to violate the pact. Since *A*'s decision is unknown to *B*, it cannot affect whether or not *B* adheres to the pact, and so the advantage gained by *A*'s violation is not outweighed by any consequent loss.

Therefore if *A* and *B* are prudent they will adhere to their disarmament pact whenever violation would be detectable by the other, and violate the pact whenever violation would not be detectable by the other. In other words, they will adhere openly and violate secretly. The disarmament pact between *A* and *B* thus possesses two of the characteristics ascribed by the thesis to morality. First, accepting the pact and acting on it is more advantageous for each than making no pact at all. Second, in so far as the pact stipulates that each must disarm even when disarming is undetectable by the other, it requires each to perform disadvantageous acts – acts which run counter to considerations of prudence.

One further condition must be met if the disarmament pact is to possess those characteristics ascribed by the thesis to a system of morality. It must be the case that the requirement that each party perform disadvantageous acts be essential to the advantage conferred by the pact; or, to put the matter in the way in which we expressed it earlier, both *A* and *B* must do better to adhere to this pact than to a pact which is similar save that it requires no disadvantageous acts. In terms of the example, *A* and *B* must do better to adhere to the pact than to a pact which stipulates that each must disarm only when disarming is detectable by the other.

We may plausibly suppose this condition to be met. Although *A* will gain by secretly retaining arms itself, it will lose by *B*'s similar acts, and its losses may well outweigh its gains. *B* may equally lose more by *A*'s secret violations than it gains by its own. So, despite the fact that prudence requires each to violate secretly, each may well do better if both adhere secretly than if both violate secretly. Supposing this to be the case, the disarmament pact is formally analogous to a moral system, as characterized by the thesis. That is, acceptance of and adherence to the pact by *A* and *B* is more advantageous for each, either than making no pact at all or than acceptance of and adherence to a pact requiring only open disarmament, and the pact requires each to perform acts of secret disarmament which are disadvantageous.

Some elementary notation, adapted for our purposes from the mathematical theory of games, may make the example even more perspicuous. Given a disarmament pact between *A* and *B*, each may pursue two pure strategies – adherence and violation. There are, then, four possible combinations of strategies, each determining a particular outcome. These outcomes can be ranked preferentially for each nation; we shall let the numerals 1 to 4 represent the ranking from first to fourth preference. Thus we construct a simple matrix,[1] in which *A*'s preferences are stated first:

		B	
		adheres	*violates*
A	adheres	2, 2	4, 1
	violates	1, 4	3, 3

The matrix does not itself show that agreement is advantageous to both, for it gives only the rankings of outcomes given the agreement. But it is plausible to assume that *A* and *B* would rank mutual violation on a par with no agreement. If we assume this, we can then indicate the value to each of making and adhering to the pact by reference to the matrix.

The matrix shows immediately that adherence to the pact is not the most advantageous possibility for either, since each prefers the outcome, if it alone violates, to the outcome of mutual adherence. It shows also that each gains less from its own violations than it loses from the other's, since each ranks mutual adherence above mutual violation.

Let us now use the matrix to show that, as we argued previously, public adherence to the pact is prudent and mutually advantageous, whereas private adherence is not prudent although mutually advantageous. Consider first the case when adherence – and so violation – are open and public.

If adherence and violation are open, then each knows the strategy chosen by the other, and can adjust its own strategy in the light of this knowledge – or, in other words, the strategies are interdependent. Suppose that each initially chooses the strategy of adherence. *A* notices that if it switches to violation it gains – moving from 2 to 1 in terms of preference ranking. Hence immediate interest dictates such a switch. But it notices further that if it switches, then *B* can also be expected to switch – moving

[1] Those familiar with the theory of games will recognize the matrix as a variant of the Prisoner's Dilemma. In a more formal treatment, it would be appropriate to develop the relation between morality and advantage by reference to the Prisoner's Dilemma. This would require reconstructing the disarmament pact and the moral system as proper games. Here I wish only to suggest the bearing of games theory on our enterprise.

from 4 to 3 on its preference scale. The eventual outcome would be stable, in that neither could benefit from switching from violation back to adherence. But the eventual outcome would represent not a gain for *A* but a loss – moving from 2 to 3 on its preference scale. Hence prudence dictates no change from the strategy of adherence. This adherence is mutually advantageous; *A* and *B* are in precisely similar positions in terms of their pact.

Consider now the case when adherence and violation are secret and private. Neither nation knows the strategy chosen by the other, so the two strategies are independent. Suppose *A* is trying to decide which strategy to follow. It does not know *B*'s choice. But it notices that if *B* adheres, then it pays *A* to violate, attaining 1 rather than 2 in terms of preference ranking. If *B* violates, then again it pays *A* to violate, attaining 3 rather than 4 on its preference scale. Hence, no matter which strategy *B* chooses, *A* will do better to violate, and so prudence dictates violation.

B of course reasons in just the same way. Hence each is moved by considerations of prudence to violate the pact, and the outcome assigns each rank 3 on its preference scale. This outcome is mutually disadvantageous to *A* and *B*, since mutual adherence would assign each rank 2 on its preference scale.

If *A* and *B* are both capable only of rational prudence, they find themselves at an impasse. The advantage of mutual adherence to the agreement when violations would be secret is not available to them, since neither can find it in his own overall interest not to violate secretly. Hence, strictly prudent nations cannot reap the maximum advantage possible from a pact of the type under examination.

Of course, what *A* and *B* will no doubt endeavour to do is eliminate the possibility of secret violations of their pact. Indeed, barring additional complications, each must find it to his advantage to make it possible for the other to detect his own violations. In other words, each must find it advantageous to ensure that their choice of strategies is interdependent, so that the pact will always be prudent for each to keep. But it may not be possible for them to ensure this, and to the extent that they cannot, prudence will prevent them from maximizing mutual advantage.

IV

We may now return to the connection of morality with advantage. Morality, if it is a system of principles of the type characterized in the thesis, requires that some persons perform acts genuinely disadvantageous to themselves, as a means to greater mutual advantage. Our example shows sufficiently that such a system is possible, and indicates more precisely its character. In particular, by an argument strictly parallel to that which we have pursued, we may show that men who are merely prudent will not perform the required disadvantageous acts. But in so violating the principles of morality, they will disadvantage themselves. Each will lose more by the violations of others than he will gain by his own violations.

Now this conclusion would be unsurprising if it were only that no man can gain if he alone is moral rather than prudent. Obviously such a man loses, for he adheres to moral principles to his own disadvantage, while others violate them also to his disadvantage. The benefit of the moral system is not one which any individual can secure for himself, since each man gains from the sacrifices of others.

What is surprising in our conclusion is that no man can ever gain if he is moral. Not only does he not gain by being moral if others are prudent, but he also does not gain by being moral if others are moral. For although he now receives the advantage of others' adherence to moral principles, he reaps the disadvantage of his own adherence. As long as his own adherence to morality is independent of what others do (and this is required to distinguish morality from prudence), he must do better to be prudent.

If all men are moral, all will do better than if all are prudent. But any one man will always do better if he is prudent than if he is moral. There is no real paradox in supposing that morality is advantageous, even though it requires the performance of disadvantageous acts.

On the supposition that morality has the characteristics ascribed to it by the thesis, is it possible to answer the question 'Why should we be moral?' where 'we' is taken distributively, so that the question is a compendious way of asking, for each person, 'Why should I be moral?' More simply, is it possible to answer the question 'Why should I be moral?'

I take it that this question, if asked seriously, demands a reason for being moral other than moral reasons themselves. It demands that moral reasons be shown to be reasons for acting by a non-circular argument. Those who would answer it, like Baier, endeavour to do so by the introduction of considerations of advantage.

Two such considerations have emerged from our discussion. The first is that if all are moral, all will do better than if all are prudent. This will serve to answer the question 'Why should we be moral?' if this question is interpreted rather as 'Why should we all be moral – rather than all being something else?' If we must all be the same, then each person has a reason – a prudential reason – to prefer that we all be moral.

But, so interpreted, 'Why should we be moral?' is not a compendious way of asking, for each person, 'Why should I be moral?' Of course, if everyone is to be whatever I am, then I should be moral. But a general answer to the question 'Why should I be moral?' cannot presuppose this.

The second consideration is that any individual always does better to be prudent rather than moral, provided his choice does not determine other choices. But in so far as this answers the question 'Why should I be moral?' it leads to the conclusion 'I should not be moral'. One feels that this is not the answer which is wanted.

We may put the matter otherwise. The individual who needs a reason for being moral which is not itself a moral reason cannot have it. There is nothing surprising about this; it would be much more surprising if such reasons could be found. For it is more than apparently paradoxical to suppose that considerations of advantage could ever of themselves justify accepting a real disadvantage.

V

I suggested in Section II that the thesis, in modified form, might provide a necessary, although not a sufficient, condition for a moral system. I want now to consider how one might characterize the man who would qualify as moral according to the thesis – I shall call him the 'moral' man – and then ask what would be lacking from this characterization, in terms of some of our commonplace moral views.

The rationally prudent man is incapable of moral behaviour, in even the limited sense defined by the thesis. What difference must there be between the prudent man and the 'moral' man? Most simply, the 'moral' man is the prudent but trustworthy man. I treat trustworthiness as the capacity which enables its possessor to adhere, and to judge that he ought to adhere, to a commitment which he has made, without regard to considerations of advantage.

The prudent but trustworthy man does not possess this capacity completely. He is capable of trustworthy behaviour only in so far as he regards his *commitment* as advantageous. Thus he differs from the prudent man just in the relevant respect; he accepts arguments of the form 'If it is advantageous for me to agree[1] to do *x*, and I do agree to do *x*, then I ought to do *x*, whether or not it then proves advantageous for me to do *x*.'

Suppose that *A* and *B*, the parties to the disarmament pact, are prudent but trustworthy. *A*, considering whether or not secretly to violate the agreement, reasons that its advantage in making and keeping the agreement, provided *B* does so as well, is greater than its advantage in not making it. If it can assume that *B* reasons in the same way, then it is in a position to conclude that it ought not to violate the pact. Although violation would be advantageous, consideration of this advantage is ruled out by *A*'s trustworthiness, given the advantage in agreeing to the pact.

The prudent but trustworthy man meets the requirements implicitly imposed by the thesis for the 'moral' man. But how

[1] The word 'agree' requires elucidation. It is essential not to confuse an advantage in agreeing to do *x* with an advantage in saying that one will do *x*. If it is advantageous for me to agree to do *x*, then there is some set of actions open to me which includes both saying that I will do *x* and doing *x*, and which is more advantageous to me than any set of actions open to me which does not include saying that I will do *x*. On the other hand, if it is advantageous for me to say that I will do *x*, then there is some set of actions open to me which includes saying that I will do *x*, and which is more advantageous to me than any set which does not include saying that I will do *x*. But this set need not include doing *x*.

far does this 'moral' man display two characteristics commonly associated with morality – first, a willingness to make sacrifices, and second, a concern with fairness?

Whenever a man ignores his own advantage for reasons other than those of greater advantage, he may be said to make some sacrifice. The 'moral' man, in being trustworthy, is thus required to make certain sacrifices. But these are extremely limited. And – not surprisingly, given the general direction of our argument – it is quite possible that they limit the advantages which the 'moral' man can secure.

Once more let us turn to our example. *A* and *B* have entered into a disarmament agreement and, being prudent but trustworthy, are faithfully carrying it out. The government of *A* is now informed by its scientists, however, that they have developed an effective missile defence, which will render *A* invulnerable to attack by any of the weapons actually or potentially at *B*'s disposal, barring unforeseen technological developments. Furthermore, this defence can be installed secretly. The government is now called upon to decide whether to violate its agreement with *B*, install the new defence, and, with the arms it has retained through its violation, establish its dominance over *B*.

A is in a type of situation quite different from that previously considered. For it is not just that *A* will do better by secretly violating its agreement. *A* reasons not only that it will do better to violate no matter what *B* does, but that it will do better if both violate than if both continue to adhere to the pact. *A* is now in a position to gain from abandoning the agreement; it no longer finds mutual adherence advantageous.

We may represent this new situation in another matrix:

		B	
		adheres	*violates*
A	adheres	3, 2	4, 1
	violates	1, 4	2, 3

We assume again that the ranking of mutual violation is the same as that of no agreement. Now had this situation obtained at the outset, no agreement would have been made, for *A* would have had no reason to enter into a disarmament pact. And of course had *A* expected this situation to come about, no agreement – or only a temporary agreement – would have been made; *A* would no doubt have risked the short-term dangers of the continuing arms race in the hope of securing the long-run benefit of predominance over *B* once its missile defence was completed. On the contrary, *A* expected to benefit from the agreement, but now finds that, because of its unexpected development of a missile defence, the agreement is not in fact advantageous to it.

The prudent but trustworthy man is willing to carry out his agreements, and judges that he ought to carry them out, in so far as he considers them advantageous. *A* is prudent but trustworthy. But is *A* willing to carry out its agreement to disarm, now that it no longer considers the agreement advantageous?

If *A* adheres to its agreement in this situation, it makes a sacrifice greater than any advantage it receives from the similar sacrifices of others. It makes a sacrifice greater in kind than any which can be required by a mutually advantageous agreement. It must, then, possess a capacity for trustworthy behaviour greater than that ascribed to the merely prudent but trustworthy man (or nation). This capacity need not be unlimited; it need not extend to a willingness to adhere to any commitment no matter what sacrifice is involved. But it must involve a willingness to adhere to a commitment made in the expectation of advantage, should that expectation be disappointed.

I shall call the man (or nation) who is willing to adhere, and judges that he ought to adhere, to his prudentially undertaken agreements even if they prove disadvantageous to him, the trustworthy man. It is likely that there are advantages available to trustworthy men which are not available to merely prudent but trustworthy men. For there may be situations in which men can make agreements which each expects to be advantageous to him, provided he can count on the others' adhering to it whether or not their expectation of advantage is realized. But each can count on this only if all have the capacity to adhere to commitments regardless of whether the commitment actually proves advantageous. Hence, only trustworthy men who know each other to be such will be able rationally to enter into, and so to benefit from, such agreements.

Baier's view of morality departs from that stated in the thesis in that it requires trustworthy, and not merely prudent but trustworthy, men. Baier admits that 'a person might do better for himself by following enlightened self-interest rather than morality'.[1] This admission seems to require that morality be a system of principles which each person may expect, initially, to be advantageous to him, if adopted and adhered to by everyone, but not a system which actually is advantageous to everyone.

Our commonplace moral views do, I think, support the view that the moral man must be trustworthy. Hence, we have established one modification required in the thesis, if it is to provide a more adequate set of conditions for a moral system.

But there is a much more basic respect in which the 'moral' man falls short of our expectations. He is willing to temper his single-minded pursuit of advantage only by accepting the obligation to adhere to prudentially undertaken commitments. He has no real concern for the advantage of others, which would lead him to modify his pursuit of advantage when it conflicted with the similar pursuits of others. Unless he expects to gain, he is unwilling to accept restrictions on the pursuit of advantage which are intended to equalize the opportunities open to all. In other words, he has no concern with fairness.

We tend to think of the moral man as one who does not seek his own well-being by means which would deny equal well-being to his fellows. This marks him off clearly from the 'moral' man, who differs from the prudent man only in that he can overcome the apparent paradox of prudence and obtain those advantages which are available only to those who can display real restraint in their pursuit of advantage.

Thus a system of principles might meet the conditions laid down in the thesis without taking any account of considerations of fairness. Such a system would contain principles for ensuring increased advantage (or expectation of advantage) to everyone, but no further principle need be present to determine the distribution of this increase.

[1] Baier, op. cit., p. 314.

It is possible that there are systems of principles which, if adopted and adhered to, provide advantages which strictly prudent men, however rational, cannot attain. These advantages are a function of the sacrifices which the principles impose on their adherents.

Morality may be such a system. If it is, this would explain our expectation that we should all be worse off were we to substitute prudence for morality in our deliberations. But to characterize morality as a system of principles advantageous to all is not to answer the question 'Why should I be moral?' nor is it to provide for those considerations of fairness which are equally essential to our moral understanding.

BIBLIOGRAPHY

We give here a list of books and articles on some of the topics discussed in the collection. The function of a bracketed number (or numbers) after an item is to indicate the topic (or topics) it covers, the key being as follows:

1. Universalizability.
2. Prescriptivity.
3. That moral rules and principles are overriding.
4. That moral rules and principles are important.
5. That moral rules and principles are associated with particular kinds of sanction.
6. That moral rules and principles have a certain content.
7. The parts of morality.
8. The justification of morality.
9. Weakness of will.
10. The derivability of evaluative from factual statements.

H. D. AIKEN, 'The concept of moral objectivity', in *Morality and the Language of Conduct*, H. N. Castañeda and G. Nakhnikian (eds.), Wayne State University, 1963 (3).

G. E. M. ANSCOMBE, 'Modern moral philosophy', *Philosophy*, 1958.

G. E. M. ANSCOMBE, 'On brute facts', *Analysis*, 1957–8 (10).

ARISTOTLE, *Nicomachean Ethics* (various translations), Book VII (9).

K. BAIER, 'The point of view of morality', *Australasian Journal of Philosophy*, 1954 (5, 6).

K. BAIER, *The Moral Point of View*, Cornell, 1958; Chapter VIII (6).

J. R. BAMBROUGH, 'The Socratic paradox', *Philosophical Quarterly*, 1960 (9).

R. W. BEARDSMORE, 'Consequences and moral worth', *Analysis*, 1968–9.

S. I. BENN and R. S. PETERS, *Social Principles and the Democratic State*, Allen and Unwin, 1959; Chapter II (6).

J. BENNETT, 'Whatever the consequences', *Analysis*, 1965–6.

J. BENNETT, 'Critical discussion of D.P. Gauthier, *Practical Reasoning*', *Mind*, 1965 (5).

M. BLACK, 'The Gap between "is" and "should" ', *Philosophical Review*, 1964 (10).

F. H. BRADLEY, *Ethical Studies*, Oxford, 1876; Essay II (8).

R. B. BRANDT, 'The definition of an "ideal observer" theory in ethics', *Philosophy and Phenomenological Research*, 1954–5.

R. N. BRONAUGH, 'Formal criteria for moral rules', *Mind*, 1968.

J. A. BRUNTON, 'Restricted moralities', *Philosophy*, 1966 (1).

J. CARGILE, 'On consequentialism', *Analysis*, 1968–9.

B. COHEN, 'An ethical paradox', *Mind*, 1967.

M. F. COHEN, ' "Is and should": an unbridged gap', *Philosophical Review*, 1965 (10).

N. COOPER, 'Some presuppositions of moral judgments', *Mind*, 1966.

N. COOPER, 'Two concepts of morality', *Philosophy*, 1966.

N. COOPER, 'Morality and importance', *Mind*, 1968.

A. DONAGAN, 'Mr Hare and the conscientious Nazi', *Philosophical Studies*, 1964–5 (2).

A. EDEL, *Method in Ethical Theory*, Routledge and Kegan Paul, 1963.

D. EMMET, 'Universalizability and moral judgment', *Philosophical Quarterly*, 1963 (1).

A. C. EWING, *Second Thoughts in Moral Philosophy*, Routledge and Kegan Paul, 1959; Chapter I (9).

W. D. FALK, 'Morality, self and others', in *Morality and the*

Language of Conduct, H.N. Castañeda and G. Nakhnikian (eds.), Wayne State University, 1963 (3).

R. FIRTH, 'Ethical absolutism and the ideal observer', *Philosophy and Phenomenological Research*, 1951–2.

A. G. N. FLEW, 'On not deriving "ought" from "is" ', *Analysis*, 1964–5 (10).

P. R. FOOT, 'When is a principle a moral principle?', *Proceedings of the Aristotelian Society*, Supplementary volume, 1954 (6).

P. R. FOOT, 'Moral beliefs', *Proceedings of the Aristotelian Society*, 1958–9 (6).

P. R. FOOT, 'Moral arguments', *Mind*, 1958 (6, 10).

W. K. FRANKENA, 'The Naturalistic Fallacy', *Mind*, 1939 (10).

W. K. FRANKENA, 'MacIntyre on defining "morality" ', *Philosophy*, 1958.

W. K. FRANKENA, 'The concept of morality', *Journal of Philosophy*, 1966.

W. K. FRANKENA, 'The concept of morality', *University of Colorado Studies*, Series in Philosophy No. 3, 1967.

P. L. GARDINER, 'On assenting to a moral principle', *Proceedings of the Aristotelian Society*, 1954–5 (2).

D. P. GAUTHIER, *Practical Reasoning*, Oxford, 1963; Chapter VI (6).

D. P. GAUTHIER, 'Morality and advantage', *Philosophical Review*, 1967.

D. P. GAUTHIER, 'Hare's debtors', *Mind*, 1968.

E. A. GELLNER, 'Ethics and logic', *Proceedings of the Aristotelian Society*, 1954–5 (1).

C. K. GRANT, 'Akrasia and the criteria of assent to practical principles', *Mind*, 1956 (9).

G. R. GRICE, *The Grounds of Moral Judgment*, Cambridge, 1967.

A. P. GRIFFITHS, 'Justifying moral principles', *Proceedings of the Aristotelian Society*, 1957–8 (8).

S. N. HAMPSHIRE, *Thought and Action*, Chatto and Windus, 1959; Chapter IV (4).

R. M. HARE, *The Language of Morals*, Oxford, 1952.

R. M. HARE, 'Universalizability', *Proceedings of the Aristotelian Society*, 1954–5 (1).

R. M. HARE, *Freedom and Reason*, Oxford, 1963.

R. M. HARE, 'Review of G. J. Warnock, *Contemporary Moral Philosophy*', *Mind*, 1968.

J. HARRISON, 'Utilitarianism, universalization and our duty to be just', *Proceedings of the Aristotelian Society*, 1952–3 (1).

J. HARRISON, 'When is a principle a moral principle?', *Proceedings of the Aristotelian Society*, Supplementary volume, 1954.

H. L. A. HART, 'Legal and moral obligation', in *Essays in Moral Philosophy*, A. I. Melden (ed.), Washington University Press, 1958.

H. L. A. HART, *The Concept of Law*, Oxford, 1961; Chapter VIII (4, 5).

T. HOBBES, *Leviathan* (1651; various editions).

H. J. N. HORSBURGH, 'The criteria of assent to a moral rule', *Mind*, 1954 (9).

H. J. N. HORSBURGH, 'The plurality of moral standards', *Philosophy*, 1954 (7).

W. D. HUDSON, 'The "is-ought" controversy', *Analysis*, 1964–5 (10).

G. E. HUGHES, 'Moral condemnation', in *Essays in Moral Philosophy*, A. I. Melden (ed.), Washington University Press, 1958.

D. HUME, *A Treatise of Human Nature* (1738; various editions).

D. HUME, *An Enquiry Concerning the Principles of Morals* (1752; various editions).

I. KANT, *Groundwork of the Metaphysic of Morals* (translated by H. J. Paton as *The Moral Law*, Hutchinson, 1948).

J. KEMP, *Reason, Action and Morality*, Routledge and Kegan Paul, 1964 (6).

G. C. KERNER, *The Revolution in Ethical Theory*, Oxford, 1966.

W. C. KNEALE, 'Moral objectivity', *Philosophy*, 1950.

J. KOVESI, *Moral Notions*, Routledge and Kegan Paul, 1967.

D. LOCKE, 'The trivializability of universalizability', *Philosophical Review*, 1968 (1).

L. M. LORING, *Two Kinds of Values*, Routledge and Kegan Paul, 1966.

S. LUKES, 'Moral weakness', *Philosophical Quarterly*, 1965 (9).

D. LYONS, *Forms and Limits of Utilitarianism*, Oxford, 1965.

H. J. MCCLOSKEY, 'Suppose everyone did the same', *Mind*, 1966.

M. C. MCGUIRE, 'Can I do what I think I ought not to? Where has Hare gone wrong?', *Mind*, 1961 (9).

A. C. MACINTYRE, 'What morality is not', *Philosophy*, 1957 (1, 2).

A. C. MACINTYRE, 'Hume on "is" and "ought" ', *Philosophical Review*, 1959 (10).

A. C. MACINTYRE, *A Short History of Ethics*, Macmillan, 1966; Chapters I and XVIII.

P. T. MACKENZIE, 'Fact and value', *Mind*, 1967 (10).

G. M. MATTHEWS, 'Evaluative and descriptive', *Mind*, 1958 (10).

G. M. MATTHEWS, 'Weakness of will', *Mind*, 1966 (9).

B. MAYO, *Ethics and the Moral Life*, Macmillan, 1958.

J. S. MILL, *Utilitarianism* (1861; various editions) (6).

D. H. MUNRO, 'Impartiality and consistency', *Philosophy*, 1961 (1).

D. H. MUNRO, 'Review of R. M. Hare, *Freedom and Reason*', *Australasian Journal of Philosophy*, 1964.

D. H. MUNRO, *Empiricism and Ethics*, Cambridge, 1967 (1, 3).

R. D. L. MONTAGUE, 'Universalizability', *Analysis*, 1964–5 (1).

R. D. L. MONTAGUE, ' "Ought" from "is" ', *American Journal of Philosophy*, 1965 (10).

G. E. MOORE, *Principia Ethica*, Cambridge, 1903.

G. E. MOORE, *Ethics*, Home University Library, 1912.

J. O. MOUNCE and D. Z. PHILLIPS, 'On morality's having a point', *Philosophy*, 1965 (1).

K. NEILSEN, 'Is "Why should I be moral?" an absurdity?', *Australasian Journal of Philosophy*, 1958 (8).

K. NEILSEN, 'On being moral', *Philosophical Studies*, 1965–6.

P. H. NOWELL-SMITH, *Ethics*, Penguin Books, 1954.

R. S. PETERS and S. I. BENN, *Social Principles and the Democratic State*, Allen and Unwin, 1959; Chapter II (6).

D. Z. PHILLIPS and J. O. MOUNCE, 'On morality's having a point', *Philosophy*, 1965 (1).

D. Z. PHILLIPS, 'Does it pay to be good?', *Proceedings of the Aristotelian Society*, 1964–5 (8).

D. Z. PHILLIPS and H. S. PRICE, 'Remorse without repudiation', *Analysis*, 1967–8.

D. Z. PHILLIPS, 'Miss Anscombe's grocer', *Analysis*, 1967–8.

H. S. PRICE and D. Z. PHILLIPS, 'Remorse without repudiation', *Analysis*, 1967–8.

A. N. PRIOR, *Logic and the Basis of Ethics*, Oxford, 1949.

A. N. PRIOR, 'The autonomy of ethics', *Australasian Journal of Philosophy*, 1960 (10).

E. G. REEVE, 'Suppose everyone did the same', *Mind*, 1969.

R. ROBINSON, 'Argument and moral argument', *Mind*, 1961 (10).

W. D. ROSS, *The Right and the Good*, Oxford, 1930.

W. D. ROSS, *Foundations of Ethics*, Oxford, 1939.

R. ROSTHAL, 'Moral weakness and remorse', *Mind*, 1967.

A. RYAN, 'Universalizability', *Analysis*, 1964–5 (1).

J. R. SEARLE, 'How to derive "ought" from "is"', *Philosophical Review*, 1964 (10).

L. A. SELBY-BIGGE, *British Moralists* (being selections from writers principally of the eighteenth century), Oxford, 1897.

H. SIDGWICK, *The Methods of Ethics*, MacMillan, 1874.

M. G. SINGER, *Generalization in Ethics*, Eyre and Spottiswoode, 1963.

F. E. SPARSHOTT, 'Critical study of R. M. Hare, *Freedom and Reason*', *Philosophical Quarterly*, 1964.

T. L. S. SPRIGGE, 'Definition of a moral judgment', *Philosophy*, 1964 (5).

C. L. STEVENSON, *Ethics and Language*, Yale University Press, 1944.

C. L. STEVENSON, *Facts and Values*, Yale University Press, 1963.

A. STIGEN, 'Mrs Foot on moral arguments', *Mind*, 1960 (10).

A. K. STOUT, 'But suppose everyone did the same', *Australasian Journal of Philosophy*, 1954.

P. F. STRAWSON, 'Social morality and individual ideal', *Philosophy*, 1961 (7).

C. C. W. TAYLOR, 'Critical notice of R. M. Hare, *Freedom and Reason*', *Mind*, 1965 (2).

I. THALBERG, 'Remorse', *Mind*, 1963.

J. and J. THOMSON, 'How not to derive "ought" from "is"', *Philosophical Review*, 1964 (10).

J. C. THORNTON, 'Can the moral point of view be justified?', *Australasian Journal of Philosophy*, 1964 (8).

S. E. TOULMIN, *The Place of Reason in Ethics*, Cambridge, 1950 (6).

J. O. URMSON, 'On grading', *Mind*, 1950 (10).

J. O. URMSON, 'Saints and heroes', in *Essays in Moral Philosophy*, A. I. Melden (ed.), Washington University Press, 1958 (7).

G. H. VON WRIGHT, *The Varieties of Goodness*, Routledge and Kegan Paul, 1963 (6).

P. S. WADIA, 'Why should I be moral?', *Australasian Journal of Philosophy*, 1964 (8).

J. J. WALSH, *Aristotle's Conception of Moral Weakness*, Columbia University Press, 1963 (9).

G. J. WARNOCK, *Contemporary Moral Philosophy*, Macmillan, 1967 (6).

G. J. WARNOCK, 'The primacy of practical reason', *Proceedings of the British Academy*, 1966 (3, 8).

G. J. WARNOCK, 'Ethics and language', *Royal Institute of Philosophy Lectures*, Series I, 1967 (6).

P. WERTHEIM, 'Morality and advantage', *Australasian Journal of Philosophy*, 1964 (8).

C. H. WHITELEY, 'On defining "moral" ', *Analysis*, 1959–60.

C. H. WHITELEY, 'Universalizability', *Analysis*, 1966–7 (1).

B. A. O. WILLIAMS, 'Ethical consistency', *Proceedings of the Aristotelian Society*, Supplementary volume, 1965.

C. WILLIAMSON, 'The grocers of Miss Anscombe and Mr Phillips', *Analysis*, 1967–8.

J. WILSON, *Reason and Morals*, Cambridge, 1961; Chapter II.

P. G. WINCH, 'The universalizability of moral judgments', *The Monist*, 1965 (1).

P. G. WINCH, 'Can a good man be harmed?', *Proceedings of the Aristotelian Society*, 1965–6 (8).

L. WITTGENSTEIN, 'Lecture on ethics', *Philosophical Review*, 1965.

CONTRIBUTORS

MISS G. E. M. ANSCOMBE is a Professor of Philosophy at Cambridge University.

KURT BAIER is a Professor of Philosophy at Pittsburgh University.

NEIL COOPER is Senior Lecturer in Philosophy at the University of Dundee.

MRS PHILIPPA FOOT is a Fellow and Tutor in Philosophy at Somerville College, Oxford.

W. K. FRANKENA is a Professor of Philosophy at the University of Michigan.

D. P. GAUTHIER is a Professor of Philosophy at the University of Toronto.

ALASDAIR MACINTYRE is a Professor in the Department of Sociology at Essex University.

T. L. S. SPRIGGE is a Lecturer in Philosophy at the University of Sussex.

P. F. STRAWSON is Wayneflete Professor of Metaphysical Philosophy at Oxford University and a Fellow of Magdalen College.

C. C. W. TAYLOR is a Fellow and Tutor in Philosophy at Corpus Christi College, Oxford.

C. H. WHITELEY is Professor of Philosophy at Birmingham University.

NAME INDEX

(not including authors mentioned only in the Bibliography)

For Product Safety Concerns and Information please contact our EU representative GPSR@taylorandfrancis.com Taylor & Francis Verlag GmbH, Kaufingerstraße 24, 80331 München, Germany

Batch number: 10725802

Printed by Printforce, the Netherlands